INFORMATION TECHNOLOGY AND PEOPLE: DESIGNING FOR THE FUTURE

EDITED BY
FRANK BLACKLER & DAVID OBORNE

DESIGNING FOR THE FUTURE

INFORMATION TECHNOLOGY

People

EDITED BY

Frank Blackler and David Oborne

THE BRITISH PSYCHOLOGICAL SOCIETY

First MIT Press paperback edition, 2003

First edition published in 1987 by The British Psychological Society, St Andrews House, 48 Princess Road East, Leicester, LEI 7DR, UK.

Available in the UK, Europe, Africa, Asia and Australasia from:
The British Psychological Society
The Distribution Centre
Blackhorse Road
LETCHWORTH
Hertfordshire
SG6 1HN
UK

Available in North and South America, Canada, the Philippines, US dependencies and Japan from:
The MIT Press

British Library Cataloguing in Publication Data

Information technology and people: designing for the future.
1. Information storage and retrieval systems
2. Man—machine systems
I. Blackler, F.H.M. II. Oborne, David III. British Psychological Society
001. 5 Z699

Library of Congress Cataloging-in-Publication Data

Information technology and people.

1. Electronic data processing—Social aspects.
I. Blackler, F. H. M. II. Oborne, David J.
QA76.9.C66I526 1987 303.4'834 86–20119
ISBN 978-0-262-02260-6 (hc : alk. paper)
978-0-262-52404-9 (pb : alk. paper)

PREFACE

For the foreseeable future, information technology is likely to be one of the most potent growth areas in advanced industrialized countries. Indeed it is now widely recognized that long-term economic prosperity will crucially depend upon people's success in developing, mastering, exploiting and marketing information systems.

Although the argument for the rapid development and introduction of information technology (IT) into the many aspects of our everyday existence is extremely strong, unfortunately it remains the case that at present the technology is being used *effectively* by only a small proportion of the people who could benefit from it. As the chapters in this book illustrate, examples of wasted resources, time and effort as a result of poor IT application or implementation are legion. Many countries and institutions are now asking why the mismatch between predicted benefits and actual performance has occured. This book suggests an approach to answering such questions.

The message of the volume is clear. It is that previous attempts to apply and implement IT into many different types of setting have been doomed from the beginning because, typically, one crucial component has been mismanaged: *people*. No matter how well-engineered IT systems (hardware, firmware, software) may be, they remain totally inanimate until people are introduced. Unless the new technologies are designed and introduced with an appreciation of the needs and reactions of those who will use and be affected by them, the chances that they will operate effectively are slight.

It is people who have to operate systems built around the new technologies, and who have to interact with them to make them function. It is people who have to accept the systems and their intrusion into work and home. It is people who are often called on to adapt their habits to the dictates of the system, and who often cannot do so. The simple answer to the questions that are currently being posed about the effectiveness of IT systems is that if people do not, will not or cannot do these things then the systems will not function adequately; indeed, they may not function at all.

In short, the application of knowledge about human behaviour, actions, thoughts, feelings, education, etc., has often been insufficient or missing. This is both unfortunate and unnecessary, for sophisticated behavioural models and methods are now available. As the chapters in this volume indicate, psychological approaches have much to offer both in predicting and understanding users' potential needs and problems, and in developing new approaches to the crucial issues of technological design and management.

Frank Blackler
David J. Oborne

CONTENTS

Part Three
SOCIETY AND THE NEW TECHNOLOGIES

PSYCHOLOGY AND INFORMATION TECHNOLOGY

Ian Howarth

BACKGROUND: THE ECONOMIC IMPORTANCE OF THE HUMAN FACTOR

There is no doubt that the economic performance of developed countries will be critically affected by their success in mastering, exploiting and marketing new developments in information technology. Our principal competitors may be advancing more rapidly than the UK in the economic exploitation of information technology. There is unfortunately very clear evidence, over the past five years, that UK manufacturers are taking a declining proportion of both home and overseas markets in IT products. This relative decline appears to have many causes, including short – sighted investment policies and too few resources habitually devoted to technical research and development by both government and industry in the UK. These problems are well known and efforts are being made to solve them. There is, however, considerably less appreciation of the greater efforts which our competitors devote to human factors research in the development of new products, and of the greater care which they take over human aspects of the exploitation of new technology. As a result their products tend to be more 'user friendly' in the sense that individuals feel more comfortable with them and organizations can integrate them more easily into their continuing activities.

With government support the Alvey research programme in information technology has stimulated new developments in design and manufacturing techniques. It has also supported some research on psychological aspects of the man–machine interface and of intelligent knowledge-based systems. Surveys of the work of the Alvey Directorate and plans for a continuation of its work (for example Executive Summary of After Alvey Workshops 6–8 January 1986) suggest that relatively more effort should be devoted to research on the psychological factors limiting the exploitation of information technology.

The prime purpose of this book is to survey what is already understood about human factors in relation to information technology. A secondary purpose is to consider how this understanding can be translated into economically effective action.

In this chapter I discuss possible strategies for exploiting what we already know and for seeking new knowledge about the human factor. Among other things, I hope to convince non-psychologists such as politicians, industrialists, civil servants and engineers that psychologists have important roles to play in the proper development and use of information technology.

Let us consider the phrase 'user friendly' more analytically. There is no doubt it has been over-used and used rather carelessly. Some would like to replace it by the single word 'usability'. Whichever term is preferred, what matter are the ideas which it summarizes. Well-designed technology should be easily understood; easy to use, with a low probability of error; easily mastered, with a minimum of training by the people who need to use it; easy to maintain and to modify; and with virtues which are so manifestly apparent that it will be relatively easy to sell. This will require not only well-designed input and output devices such as keyboards, lightpens, visual displays and robot arms, but also an easily understandable 'logic' in the working of the total system.

That is a formidable list of virtues to be summarized by the single word 'usability' or the single phrase 'user friendly'. But all these virtues are likely to follow if the 'human factor' has been taken into account in the design and use of the technology. When our competitors' products are more attractive, easier to use, more adaptable, less disruptive of other activities, easier to maintain, or, in a word, more marketable than our own, it will be because they have been more successful than ourselves in their use of psychological knowledge and in evaluating the behavioural and organizational effects of their designs.

The application of psychological techniques in the development and use of information technologies is the subject matter of later chapters. In this chapter I consider various ways in which technology may be made more or less user friendly. These range from learning by experience, in a rather amateur fashion, to employing professionally trained psychologists. In this country we have tended to distrust professionals, whether they be scientists, engineers or economists. Fortunately the cult of the amateur is less bigoted than in the past and there is an increasing understanding of the need for adequately trained scientists and engineers in technically innovative industrial developments. However, a surprising number of people, including many scientists, engineers and managers still adopt an amateurish approach to human factors. As long as this attitude continues, our competitors are likely to produce products which are more 'user friendly' and hence more successful than our own.

How, then, should we approach the human factors problem posed by information technology?

LEARN FROM EXPERIENCE

This is a very common strategy. It is like the approach to engineering of the builders of medieval cathedrals. When a cathedral fell down, as they quite frequently did, the next cathedral was made slightly more sturdy. When one stood for a hundred years, the builders attempted something a little more adventurous and delicately soaring. The same approach to bridge building was adopted between Roman times and the eighteenth century. Eventually, Newtonian mathematics was applied to the design of load-bearing structures and it was recognized that professional civil engineers were the best people to do the necessary calculations. As a result, the design of buildings and bridges has changed rapidly and with relatively few disasters.

That comparison is not quite fair since we all learn much more by experience about our own humanity and the humanity of other people than we do about the load-bearing character of different physical structures. We are all amateur psychologists during most of our waking lives, but amateur engineers only rarely. However, no matter how good we become at understanding people in their habitual environments, experience is not so good a guide when we seek to predict how people will behave in novel environments.

The speed of developments in information technology produces new environments, particularly new working environments, more rapidly than at any period in our history. The changes are so rapid that experience is a fallible guide and like the medieval masons we have had our disasters. We shall of course learn from them.

The Swansea Licensing Centre for motor vehicle driver licences was intended to simplify the issuing of driver licences, since all records would be kept in a single computer. This would speed up the process of checking qualifications and reduce the number of errors in the previous system, which depended on meticulous filing of hand-written records. What happened when the Centre opened was almost a fiasco. The delays in issuing licences grew so long that the licensing system became far less reliable than it had been previously. This was not due to any deficiencies in the computer system. The problems were entirely due to the difficulties the staff had in coming to terms with the machine.

Gradually the Licensing Centre has sorted itself out. The staff have learned to cope with the system. The procedures have been simplified. We all get licences valid for 20 years or more and the system now works reasonably well. That experience will help us avoid similar difficulties when other national data bases, such as that for income tax, are set up; but the degree of help will depend on their similarity to the Swansea system and how well we have understood what went wrong. Unfortunately, although well known, the Swansea experience was not well documented and so we may have learned less than we otherwise might have done from it.

The accident at the Three Mile Island nuclear power plant in Pennsylvania was another human factors disaster which could very easily have been much worse. The fault which developed in the system was not very serious and could easily have been rectified if the men in the control room had understood it; but confronted with a vast array of information technology, they misunderstood the fault and by misunderstanding it made it infinitely worse. This particular disaster, unlike that at Swansea, has been very carefully analysed by psychologists, who have made recommendations which, we hope, will very much reduce the likelihood of similar misunderstandings in the future. The control room at Three Mile Island was far from 'user friendly'. This was made abundantly clear in the Kemeny Commission report which has been used and commented upon by many psychologists, for example Reason (in press).

In the past two years many small computer companies have either gone out of business or have been in great financial difficulties. This has happened even to some companies whose products were electronically superb. The usual explanation for their difficulties is that they mismanaged the economic side of their business, running into cash flow problems by too rapid expansion, paying too much attention to technical developments and not enough to financial control. An alternative explanation is that they did not understand their customers' needs well enough. Too many computers were over-sold on the performance which their designers and other experts could get out of them. But they were not sufficiently 'user friendly' to be of continued value to their novice customers once the fashion for computer games had run its course. The small computer companies which are now flourishing are those which learned from the mistakes of their rivals, and have, for example, produced more user friendly word processing packages (such as Amstrad) in addition to having more efficient financial management.

The learning from experience in all these cases is not much more impressive than that of the medieval masons when a flying buttress collapsed. We should be able to do better.

ELIMINATE THE HUMAN FACTOR

The next most popular approach is to regard the human factor as undesirable and to try to design it out as far as possible: hence automated offices and automated factories. Sometimes this approach is successful and the automated process is more reliable and more economical than the human process it replaced. Computer scientists, specialists in artificial intelligence and robotics, systems analysts and a whole host of specialists in new disciplines are devoted to this task of eliminating human error by removing the human from the workplace. These are, however, the very people who are creating some of the disasters from which we are so laboriously learning by experience. There must be something wrong with this approach and it is not difficult to see what it is.

So long as machines are the servants of man and not vice versa, it is, in principle, impossible to eliminate the human entirely. As machines become more intelligent, the problems of making them user friendly will become more acute rather than less so. In the meantime there are very many human processes which cannot as yet be automated and the need for machines to work with and alongside human workers is very obvious. The prime examples are perception (visual and speech perception), speech production and flexible or creative problem solving. People are needed to solve simple perceptual tasks which are still beyond the capacity of a machine. Post office workers, rather than machines, still read our postal codes, although machines do the sorting once the workers have pressed the appropriate buttons. Secretaries still interpret people's speech before putting it into a word processor. People still monitor the performance of even the cleverest machines and decide to switch them off or repair them when they, the people, decide the machines are not working properly.

The strategy of reducing the number of people required to do a job by automating some of their tasks is not desirable in itself. It is only desirable in so far as it reduces the cost of production. There are some human functions which cannot be automated and some which can only be automated by using excessively powerful and expensive computers. Hence the most economic production is likely to be achieved by using humans and machines in an optimum combination, so that all play their appropriate parts doing what they do best.

The strategy of reducing the number of people involved in work requires us to pay *more*, not less, attention to human factors. People cannot be eliminated, and the more unusual the working environment, the greater the problems of human factors design.

PSYCHOLOGY IS NO MORE THAN COMMON SENSE

An almost equally common strategy is based on the attitude that, yes, we need to pay more attention to human factors, but psychology is no more than common sense hiding behind some ugly jargon; hence we can take care of human factors problems without involving those psychologists. Like the other strategies, there is something to be said for this one; but not much. No one would deny that British industry needs to pay more attention, at all levels, to human factors problems. However, the strategy of do-it-yourself psychology is based on a misunderstanding of what professional psychology has to offer and an overestimation of the power of common sense.

The deficiences of common sense are now fairly well documented. We know that common sense is more useful in creating an illusion of understanding after the event than in making predictions before the event. Proverbs come in pairs: 'too many cooks spoil the broth', 'many hands make light work'; or 'look before you leap', 'he who hesitates is lost'. At the level of proverbial wisdom, common sense can explain anything and predict nothing.

Despite these difficulties, common sense is a useful guide in most everyday activities. It is only deficient in relation to rare or novel circumstances. It can be relied upon when our reactions are unthinking and automatic. It is a poor guide when we need to take thought.

In relation to the human factors problems which are posed by information technology, common sense is not only of little help in suggesting solutions, it is equally poor at defining the problems. Occupational psychologists, who are the psychologists most frequently consulted by industry or commerce, have a poor view of their clients' ability to diagnose the nature of a problem. Most clients have a very poor vocabulary for describing the nature of human factors problems. As a result, they usually describe the problem in terms of what they imagine to be the most likely solution to it, for example 'We have a training problem'; 'We have a selection problem'; 'Our workers are not sufficiently intelligent or motivated'; 'These displays are too difficult to read quickly'; and so on. Since every human factors problem can, in principle, be solved by improved design of equipment, or by better selection of workers, or by better training, or by better organization, or by some combination of all of these, it is never appropriate to beg the question by describing a problem in terms of its potential solution. Ideas about the most efficient solution should come after the problem has been investigated and not before.

It is not surprising that common sense is so deficient in analytical concepts and vocabulary. Common sense is a guide to action. It provides a set of reasonably reliable responses to common situations. It is understandably deficient in analysing uncommon ones. Donald Broadbent (Broadbent, Fitzgerald and Broadbent, 1986) has done some very interesting experiments which throw light on this. He has shown that people can learn to operate fairly complex computer games, such as a business game, without being able to verbalize their understanding of the rules by which they run. In other words, the wisdom we acquire by experience gives us intuitive guides to action, rather than the analytic concepts with which to describe and think about the nature of a complex task.

It is not difficult to find real-life examples of the inadequacy of the common sense approach. To take one example from a recent development in word processing. The Macintosh microcomputer developed by Apple Computers Inc. is one of the most 'user friendly' ever devised. However, some of the Macintosh software has deficiences due to the failure of its designers to take account of the structure of human problem solving strategies, which usually work from the particular to the general (bottom up) rather than the other way round (top down). Microsoft Word is a screen-based word processor. It makes full use of the sophisticated Mac interface facilities and when it first appeared it was a long way ahead of its competitors. However, it is unlikely to be a top seller because of the design fault just mentioned (Thomas Green, personal communication). There are invisible formatting characters inside the text. If the user makes an overall formatting change, such as asking for the justification of the right-hand margin, then the system makes sensible but unpredictable alterations in the invisible characters, so that, for example, the positioning of equations or headings in the centre of a page may be altered and these changes are then difficult to adjust. In contrast, MacDraw, a Macintosh graphics programme, allows the user to make trial and error corrections of elements in the program, without altering its other features. Similar errors have been made in the design of data base software and in integrated office systems. These errors could have been avoided by the application of comparatively simple psychological principles. When stated these may seem mere common sense, but only after the event.

In general, and rather ironically, the military pay more attention to human factors than does civilian industry, but even the military have experienced numerous human factors foul-ups, for example in the monitoring and planning of the Vietnam war, and in trials of new technologies in which the commanders' strategies have been determined by non-existent threats detected by radar and other remote sensors. Many failures of this type are blamed on the person operating the equipment. 'Pilot error' is often invoked when the most obvious failure is one of design.

To show that psychology transcends common sense in being able to make predictions about the future as well as explain the past, I will now predict a future difficulty before it occurs. Many applications of information technology are dependent upon the development of expert systems. These systems, which are a dominant technique in 'knowledge engineering', are intended to replace a particular form of expertise, for example in medical diagnosis, or in computer-aided design, or in the control of a chemical plant, usually by a series of if/then statements known as a 'production system'. There are two unsolved problems which limit the usefulness of expert systems. These are:

1. One of the techniques used to aid the construction of an expert system is 'knowledge elicitation', that is the self-characterization, by human experts, of the nature of this expertise. This often takes the form of decision rules which can be incorporated into a production system; but many of the conscious rules used by human experts only work because they are embedded in a framework of tacit knowledge which the expert may not be able to verbalize (Broadbent, Fitzgerald and Broadbent, 1986). Hence 'knowledge elicitation' may not be such a short cut to the development of expert systems as many people now imagine.

2. Simple production systems are relatively easy to run, since there may be only a limited number of ways in which one can procede through the sequence of if/then statements. However, before they become useful, they become large and unwieldy. They can only be controlled by sophisticated planning systems which are not yet so well understood. This limits the size and power of expert systems.

These difficulties are unlikely to be completely solved in the near future. Hence, expert systems will not do some of the things which people currently expect of them. However, there are many things which they can do, provided we understand their limitations and the type of human activity which they can replace.

WRITE A HANDBOOK OF HUMAN ENGINEERING

This is a good idea, the best yet. Moreover, it has already been done several times, most recently by three psychologists, Card, Moran and Newall (1983), who summarize very cleverly all the features of human perception, motor skill and problem solving which they consider essential for the design of word-processing systems. This book is very much appreciated by designers, since it attempts to summarize its data in a form which engineers find congenial. However, it does have some deficiencies which those using it (and any other human factors handbook) must be aware of and take into account.

The first is that the amount of psychological knowledge summarized is extremely small, but the application of this knowledge is extremely complex. The knowledge is summarized in 10 principles of operation of the human information processing system. These are stated on *one page* of the book. Seventy-eight pages are required to justify and explain these principles, while 332 pages are needed to explain their application to the design of text editors. This is not quite as bad as it sounds, because the techniques of task analysis and problem solving which are used can be generalized to other issues in information technology. Nevertheless it is impossible to take any single piece of information from the book and apply it to a design problem without being aware of the techniques and skills needed to apply that information successfully.

Or, to put it another way, such a handbook is likely to be much more useful to an experienced psychologist than to an engineer who is less well trained and less experienced in human factors work.

The second deficiency in the book by Card and colleagues is that it contains very little on problem solving or learning and nothing on psycholinguistics, or instruction, or explanation, or individual differences or social factors. In other words, there is a great deal of very important psychology which is not covered. A more adequate human factors handbook would have to be very large indeed.

USE HUMAN FACTORS STANDARDS

This is another good idea. If, during the design of a product, there is a shortage of good human factors or psychological expertise, then the design engineers can be helped by giving them certain psychological standards which the product must meet. These would be comparable to the standards of physical performance which are in the specification given to any design team. Just as the physical standards help to ensure that the technology will work in its physical environment, so the human factors standards could help to ensure that it will operate satisfactorily in its human environment.

Human factors standards are most familiar in relation to product safety. It is ethically undesirable to damage the human user by excessive radiation, or by poisoning or by creating eye strain or backache. Hence the need for health and safety standards. The same approach can be used in relation to economic performance. It is economically undesirable to create difficulties for the human user because the visual displays are difficult to interpret or because the logic of the system is difficult to understand. So standards which demand clear displays, transparent logic and easily learned operations could have very beneficial economic effects.

Unfortunately, it is impossible to solve all human factors problems by the use of a small set of design requirements. The reason for this is fairly obvious, since the design of a system, including the form of its interaction wih human users, must depend on the nature of the task and on the relative roles given to person and machine in performing the task. The standards must be adapted to the circumstances.

We might be able to overcome this difficulty by producing long lists of desirable design characteristics for most imaginable tasks and strategies for performing the tasks. This has never been attempted in a civilian context, because it is a mammoth task, and, even so, works only in relation to highly predictable developments in technology. This approach has, however, been used in a limited range of military contexts.

An alternative approach would be to develop ways of adapting the standards to the needs of the situation. This is the method which has been recommended in discussions with the Alvey Directorate. What follows is based on the results of a workshop on human factors standards sponsored by the Alvey Directorate.

It is helpful to distinguish three types of human factors standard: product, performance and procedural.

Product standards like those already described, specify concrete characteristics which are desirable. These may be used by designers as an aid to design, or by buyers when comparing the characteristics of alternative products.

Performance standards may be used when product standards are not available. Instead of specifying the characteristics of the product, they specify performance characteristics such as speed of operation, ease of learning or flexibility of use, which must be achieved when the technology is used by the kind of people who will be expected to use it when it goes into operation. Performance standards may be used by designers when assessing the performance of a prototype or by buyers when choosing between products. Since both product and performance standards must be appropriate to the task, to the technology and to whatever strategic mixture of human and technical intelligence is adopted, we need a third type of standard to ensure that both product and performance standards are properly developed and evaluated.

Procedural standards specify desirable or essential features of the procedures used by the designers in developing products. They will include such things as the need for a proper task analysis before specifying what the new technology will be asked to do, the development of prototypes whose performance can be assessed, the production of effective documentation and careful introduction of the product to the market with follow-up and evaluation.

Figure 1 shows one procedural sequence which ought to be specified as a procedural standard. It is a description of the design process, loosely based on Christopher Jones (1970). It describes a design sequence which can start from an appreciation of commercial pressures and proceed in a systematic fashion to respond to these pressures. A procedural standard of this kind is not a specification of a formal system, but an *aide-mémoire* to ensure that no essential features of the design process are omitted.

The sequence of activities represented in Figure 1 can be regarded as the result of a task analysis on a well-conducted design process and

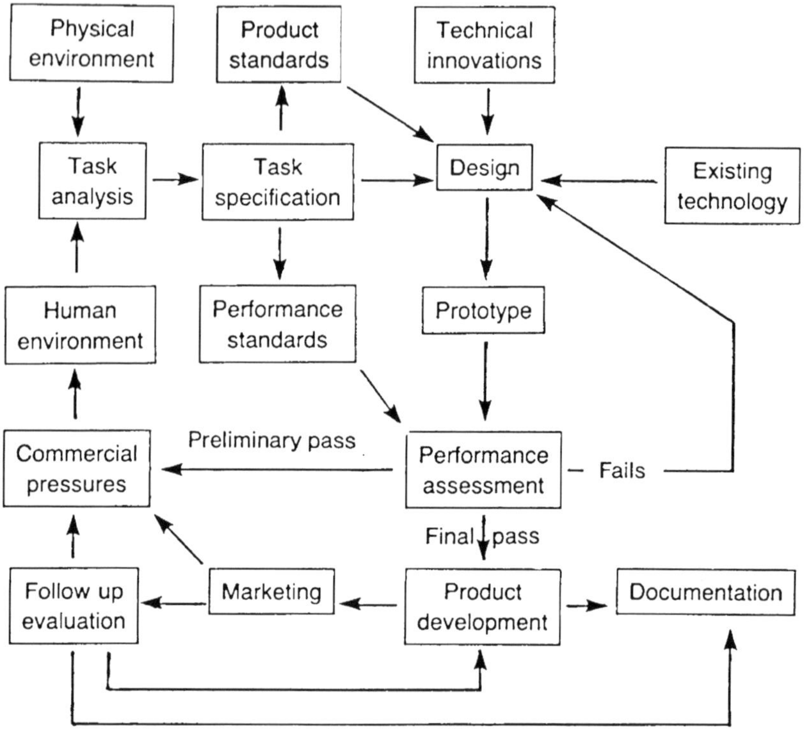

Figure 1. The design process (based on Jones, 1970).

as an *aide-mémoire* or heuristic which will help the designer to consider all important aspects of the design process.

This view of design shows its relationship to the total process of developing and marketing a new product. It has the advantage of showing the related roles of different kinds of specialist, including management (involved in marketing and assessing commercial pressures), engineers (involved in prototype and product development and the deployment of existing technology), scientists (introducing technical innovations) and psychologists (involved in task analysis and performance assessment). Most of the activities shown in Figure 1 involve the collaboration of more than one specialist.

There are several loops in the process which may receive too little attention. For example, one may go straight from *commercial pressures* to *task specification*, to allow *design* to be dominated by *technical innovation*, or skimp the process of *performance assessment* on a *prototype*. These are all false economies.

Many developments are driven by new technology rather than by an appreciation of commercial opportunity. This does not require any reorganization of Figure 1 since, no matter how innovative the technology, it must be fitted into an economic niche if it is to be successful.

The procedures set out in Figure 1 are of course a considerable simplification of the design process. The cycle – *design* → *prototype* → *performance assessment* → *design* – may be traversed several times. Initially the prototype may exist only in the designer's head, then in outline form on paper, then in several versions of a partially implemented simulation or prototype, before the final version is built and tested. The earlier, cheaper iterations ensure that there are fewer problems with the final version.

Figure 1 shows both product standards and performance standards arising out of the task requirements, which in turn are derived from an analysis of the task to be undertaken. One of the dangers of adopting ready-made product or performance standards is that they may encourage the designers to spend too little time and effort in the task analysis on which the task specification and appropriate product and performance standards should depend.

TRAIN ENGINEERS TO UNDERSTAND HUMAN FACTORS

A handbook of human engineering or the specification of human factors standards will, in themselves, give engineers a better understanding of human factors. But, as my discussion of these two strategies shows, it is very difficult for engineers to make full and effective use of either, because their training does not prepare them to do so. A recent survey of courses related to information technology (for example computer science or electrical engineering) has shown that very few of them contain much study of human behaviour. As one of the respondents said, the problems raised (by IT) are dictated 'solely by the physics of the situation'. The view of many physical scientists that all technical problems can be solved by the methods of physical science is not only a symptom of the cultural disability which inhibits our capacity to take economic advantage of our scientific achievements. It is also a powerful barrier to the development of a better understanding of human factors problems, either by education or as a result of experience.

This barrier could be overcome and indeed it must be overcome. But we must not overestimate what could be achieved by inserting more teaching of human factors and more study of human behaviour into IT training and education. It would and should give engineers an understanding of the nature of human factors problems, the methods which can be used to investigate them and the types of solution which are available. It would enable them to make better use of human engineering handbooks or of human factors standards; but it would not make them experts at investigating and solving human factors problems. Psychology and ergonomics are full three-year courses at undergraduate level and postgraduate training is considered essential for most jobs in professional psychology. To expect engineers to act like professional psychologists after taking psychology as a very minor part of their professional training is frankly dotty. It is even less sensible than it would be to expect psychologists to act like professional engineers after taking electrical engineering as a minority element in their professional training. A psychologist with a background in mathematics and science could, in fact, make reasonable use of a course in electrical engineering, at least enough to make good use of the available handbooks and physical design standards. The information in physical engineering handbooks is less context-specific than that found in human engineering handbooks, and physical design standards do not usually need to be changed for every small change in the nature of the task or the type of design chosen.

Despite the comparative ease with which one can gain access to design specifications in engineering, these specifications are more effectively used by properly trained engineers than they would be by someone with only a sketchy training. The case for employing people who are properly trained in applied psychology is even stronger.

ENCOURAGE COLLABORATION BETWEEN ENGINEERS AND USERS

While psychologists and ergonomists may claim to understand the human factors involved in making optimum use of information technology, the users of IT have an even more intimate understanding, even though their understanding may be less articulate or analytic. Many people think that by putting engineers directly in touch with the users of technology, we can cut out the middleman, in this case the psychologist.

There is no doubt that engineers can become more realistic and may even be inspired to produce more creative and effective technology as a result of direct contact with the users. It is a highly desirable form of training or experience for any engineer. But again we should not expect to solve all our problems this way. The users of technology suffer from three deficiencies as informants about the human uses of information technology:

- ☐ their experience is with existing technology, not with new inventions
- ☐ they have difficulty describing simply what they have learned from their experience because they lack the necessary vocabulary and concepts
- ☐ their testimony is distorted by the well-known effects of expectation, prejudice and motivation on perception and memory.

Engineers, even when given some training in psychology, suffer from related deficiencies as elicitators of information from users. These are:

- ☐ they lack the theoretical understanding needed to extrapolate from present experience to future possibilities
- ☐ they will waste a great deal of time because they do not have experience of techniques for extracting information from informants
- ☐ they do not know how to extract reliable information from several unreliable sources using the techniques which Newell (1972) called 'converging operations'. This involves the use of very

different techniques such as observation, interviews and diaries to disentangle valid from ephemeral testimony
- ☐ they are poorly informed on techniques for studying 'statistical' effects.

These deficiencies create a mismatch between the type of knowledge possessed by users and the ability of engineers to take advantage of it. It is precisely because psychologists and ergonomists have a theoretical understanding of human behaviour and experience in techniques for improving their understanding that they are desirable members of any design team. Middlemen they may be, but it is dangerous to eliminate them. This leads us to the final strategy, which we regard as the most effective.

EMPLOY PSYCHOLOGISTS AND ERGONOMISTS IN DESIGN AND EVALUATION

Figure 1 shows where psychologists can contribute to design and development. They are better equipped than anyone else to do task analyses before task specification; to develop human factors standards for products and performance; to do performance assessment on the prototype; to develop documentation; and to do follow up evaluation on the final product. They are very valuable members of any design team and can contribute usefully to the development of marketing strategies.

Thirty years ago, I was a very junior member of the 'High Performance Research Team' at the Institute of Aviation Medicine, Farnborough. Our task was to investigate the human problems of flying a new generation of high performance aircraft. These imposed greater physiological strain on pilots, but they created even more difficulties of a psychological kind because of the reduction of the time available for many crucial decisions and the complexity of the new instrumentation. As an afterthought, our team was expected to study these problems and suggest ways of overcoming them. We were able to suggest some remedies, such as the redesign of some instruments and the use of auditory rather than visual warnings for some purposes; but most of the ideas we developed could not be implemented because they were difficult to fit into the existing structure. The strongest impression I gained from this experience was the sheer folly of treating the human factor as a 'bolt-on goody'; something which could be done after the engineering problems had been solved. It is rather dispiriting to find myself still preaching the same sermon to the unconverted or only partially converted.

Ergonomists and psychologists are now involved in the design of aircraft, if only to help meet safety requirements; but whenever a new technology is developed, we seem to slip back into old habits and fail to involve psychologists until there are a number of embarrassing human factors errors. An exception to this rule was the American programme to put men on the moon. NASA employed large numbers of psychologists in the selection and training of astronauts and in the design of the lunar modules and the lunar module simulators. As a result, the astronauts were able to operate in a totally novel environment without any serious human error. There have, of course, been human errors in the space programme, but these have involved controllers and administrators rather than the astronauts themselves during flight.

The space programme is sometimes justified because of the technological 'fall-out' which has benefitted other aspects of industry (non-stick pans!). In my view the most useful lesson is the need to involve psychologists in any ambitious development programme.

This chapter began by claiming that our economic competitors were better than us at solving human factors problems. The Americans, the Japanese, the Germans, the Scandinavians and the Dutch are all more likely than us to involve psychologists in the design process. If the arguments presented in this chapter are valid, this is not an unimportant observation. Their better use of applied psychology is likely to be a crucial factor in their better economic performance.

WHAT KINDS OF PSYCHOLOGIST ARE EQUIPPED TO WORK ON IT?

Not all psychologists are, of course, trained to work on problems relevant to information technology, but a surprising number are. All psychologists will have received a basic training in cognitive psychology, which is now regarded as the intellectual core of most degrees in psychology. Moreover, just as engineers receive some training in psychological issues, many psychologists are taught some computing and some aspects of artificial intelligence. Of particular interest here are those who have taken joint degrees with computer science or cognitive science.

Many psychologists, after their first degree, go on to do research in cognitive psychology and much of this research is relevant to information technology, being concerned with such topics as the limitations of human attention, the structure and limits of memory, strategies in human problem solving, visual search, manual skills, perception and recognition, learning and instruction. Research in social psychology may also be relevant when it is concerned with such topics as reactions to innovation and change, negotiations, the social organization of work and socio-technical interactions. The study of stress by physiological psychologists may also be relevant, since many new technologies do create stress and the safest ways to counteract stress are psychological rather than medical or pharmacological.

Other psychologists specialize in the psychology of the work-place and act as consultants as well as researchers in this field. Occupational psychologists and ergonomists are trained in very similar ways and do very similar work. The ergonomist may know a little more about anatomy and physiology and the psychologist a little more about social factors, but these differences are very slight. These people are particularly appropriate to act as members of a design team, as described in Figure 1. They will be well trained in the relevant techniques of task analysis and in methods of evaluation, either of prototypes or of final products. They will also be knowledgeable about selection, training, organizational aspects of work and basic equipment design. Being broadly trained in dealing with applied problems, they are the most appropriate people to advise on the nature of the human factors problems likely to be encountered in the design and deployment of intelligent information systems.

It is not difficult to find out what sort of help designers can get from human factors experts. There are now many books on the psychology of human computer interaction which are more extended in scope than the handbook written by Card and colleagues (1983). This book is itself one of these. Smith and Green (1980) and Weinberg (1971) are earlier examples.

SUMMARY

To be efficient and economically successful, information technology must be acceptable to the people who will use it. It must be easy to use, and be understandable rather than mysterious, so that it can be used flexibly and reliably and be easy to maintain. It must facilitate the development of good working practices and organizational structures. In other words, information technology must be 'usable' or 'user friendly'.

These virtues will not be achieved unless proper attention is paid to the human factor in implementing new technology. If this is to be done in an efficient and professional manner, psychologists and other human factors experts must take part in the design process. When engineers design for engineers it is not surprising that other people have difficulty in working with their products. And selling them.

REFERENCES

Broadbent D.E., Fitzgerald, P. and Broadbent M.H. (1986) Implicit and explicit knowledge in the control of complex systems. *British Journal of Psychology, 77*, 33–50.

Card, S.K., Moran, T.P. and Newell, A. (1983) *The Psychology of Human – Computer Interaction*. Hillsdale, N.J.: Lawrence Erlbaum Associates.

Jones, J.C. (1970) *Design Methods: Seeds of Human Futures*. Chichester: Wiley.

Newell, A. (1973) You can't play twenty questions with nature and win. In W.G. Chase (ed.) *Visual Information Processing*. New York: Academic Press.

Reason, J. (in press) Catastrophic combinations in trivial errors. In T. Cox and S. Cox (eds) *Psychology of Occupational Safety and Accident*. Chichester: Wiley.

Smith, H.T. and Green, T.R.G. (1980) *Human Interaction with Computers*. London: Academic Press.

Thomas, J.C. and Schneider, M.L. (eds) (1984) *Human Factors in Computer Systems*. New Jersey: Ablex.

Weinberg, G.M. (1971) *The Psychology of Computer Programming*. New York: Van Nostrand Reinhold.

Part 1
INFORMATION TECHNOLOGY AND ORGANIZATIONS

In many respects, the term 'new' information technology is something of a misnomer. In many larger organizations the technology has been available for some time, and its potential impact has been described by a number of authors. With the radical reduction in the price of information technology that has occurred over the past decade, however, the range of effects of the technology for groups and organizations of all sizes has been magnified. IT is reaching an ever increasing number of people, and activities and organizations are being affected in ways that were unthought of a decade or so ago. The four chapters in this section illustrate some of these effects and the ways in which psychology has been, and can be, utilized to enhance the positive and ameliorate the negative effects.

In the first chapter of this section, Blackler and Brown consider how microelectronic technologies can be integrated within modern organizational structures, and their effects on both the management and the workers. 'It is a mistake', they argue, 'to assume that the mere purchase and commission of the new technologies will automatically enhance effectiveness.' The technology needs to be understood, they argue, not only in terms of its performance specifications, but also in terms of the pattern of social relationships it imposes. Using approaches from organizational psychology, they consider the ways in which the technologies may change the structure of the organization itself and how these changes can be exploited. Adequate planning for the technologies and their effects requires an understanding both of the alternative ways in which they can be used and of the organizational constraints that may inhibit innovative approaches.

Clegg and Wall apply ideas specifically to manufacturing industry. They argue that the choices facing management are complex. An understanding of the social and organizational consequences of automation is, however, essential if the new technologies are to contribute to an improvement in industrial competitiveness.

Oborne focuses on the extent to which the new technologies can be used effectively by end-users. Ergonomics seeks to enhance the link between people and their environment by understanding the behaviour of people so that the environment can be designed to 'fit' them. Using various examples from the new technology, he illustrates the need to consider all aspects of the operator's behaviour – cognitive, physical and physiological – when interacting with computer systems. He introduces the concept of a symbiotic relationship between the operator and his or her environment – each, without the other, cannot function adequately.

Finally in this section, Christie and Gardiner consider the specific organizational environment of the office. In recent years it is this area that has probably received most attention from the information technology manufacturers. This has occurred for a number of reasons: offices generally deal mainly with information; much of the work is apparently routine; and it is an area in which one individual (for example a secretary) often performs many different tasks. However, as Christie and Gardiner point out, the information technology impact within modern offices will not be in any way complete unless the designers and manufacturers of IT hardware and software understand why and how various office practices are carried out. In many respects, the theme which runs through all four chapters within this section is emphasized most specifically in this chapter – both the functionality and the usability of any piece of IT from the point of view of the user needs to be understood before it will either be marketed or used effectively.

MANAGEMENT, ORGANIZATIONS AND THE NEW TECHNOLOGIES

Frank Blackler and Colin Brown

In this chapter approaches from organizational psychology to the effective use of microelectronic technologies within work organizations are discussed.

First, approaches to understanding the effects of IT on organizations are considered. Information technologies are especially versatile in their applications. There is no simple cause and effect relationship between their adoption and the structural and behavioural consequences that they may have.

Second, an analysis is presented to help people recognize the organizational options raised by IT. It is important not to think of technology in terms of 'nuts and bolts' alone. Rather, in any particular work situation information technology should be thought of as a pattern of social relationships.

Third, ways of developing the general policy options that are raised by the technology are described. Within business organizations a variety of pressures typically exists that may encourage conventional approaches towards the use of IT. New approaches to the design process are needed to enable innovative approaches to the use of the technology to be developed.

Fourth, ways in which innovative changes may be introduced into organizations are outlined. Resistance to change can build up within work organizations. But, handled effectively, the introduction of IT can lead to significant long-term improvements in organizational effectiveness.

UNDERSTANDING THE EFFECTS OF INFORMATION TECHNOLOGY ON ORGANIZATIONS

The benefits of IT for work organizations

Many different products have been developed which exploit the unique features of microelectronics to capture, store, manipulate, use, present and distribute information. These include, in manufacturing industry, computer numerically controlled tools, robotics and flexible manufacturing systems, computer-aided design, possibly integrated with computer-aided manufacturing, and process control machinery. Such equipment enables faster, more reliable and accurate control of functions and a unification of previously discrete operations. In office and administrative work, applications include word processors, electronic mail and filing systems, information and decision support systems, knowledge-based programmes, electronic point-of-sale systems, electronic funds transfer and, linked to the new communications technologies, international information networks. As with the new manufacturing technologies, advanced office systems permit a fast and accurate processing of information. They can also enable effective communication links to be built between dispersed units and, because of the structured and unambiguous nature of the data accepted by computing systems, can standardize approaches to decision making and reduce the scope for professional mystique.

Numerous writers have argued that early adoption of these technologies is essential if the United Kingdom is to survive as a trading nation (for example, ACARD, 1979). Broadly speaking, work systems based on microelectronics offer two main classes of benefit. First, automation may enable significant reductions in overheads to be obtained by, for example, reductions in staff. Second are the 'value added' gains that can be obtained. These can include improved quality, design, speed of response and reliability. Thus, information technologies can be used as part of a long-term business strategy, to improve an organization's market share or to develop a reputation for quality services and customer satisfaction. It may not be possible to utilize short-term return on investment calculations to justify the introduction of new work systems intended to maximize 'value added' benefits. But while possible 'cost reduction' benefits tend to be overestimated by some suppliers of the new equipment, the most important benefits of information technologies are likely to occur when the technologies are introduced as part of a long-term business strategy.

The effects of IT on organizational structures and employee behaviour

Closely linked to a strategic use of the technologies are the organizational possibilities they raise. To obtain full benefit from many of the advanced manufacturing and office technologies, changes in an organization's arrangements may be necessary. The technology offers unprecedented opportunities to simplify organizations. This can be achieved in a number of ways and many commentators have attempted to predict the pattern of likely trends. The style of early predictions in this area was set by Leavitt and Whisler (1958) in their attempt to forecast the impact that the computing technologies of the late 1950s would have on the organization of work in future decades. They took the view that the nature of the advanced computing technology of that time would lead to an increased centralization of decision making, with a decline in the numbers of middle managers, the emergence of a powerful senior management élite and an increasing vertical segregation. Similar predictions have been made more recently. Arguing from an analysis of managerial needs within the political and economic conditions of the day, Child (1984) suggests that a trend towards greater centralization is indeed now beginning to occur. Modern computing technology offers, he suggests, attractive opportunities for the development of systems designed to enable organizations to be managed by small, flexible and centralized management élites. Arguing from an analysis of pressures within capitalist organizations, Cooley (1981) has suggested similar developments at the level of job design. Microelectronic technologies will, he argues, herald a new era of deskilling as control over discretionary elements in work is passed to machines.

Such attempts to predict the organizational impacts of the technologies involve an emphasis on certain of their possible features and the assumption either that such capabilities are so impressive that efforts will inevitably be made to exploit them, or that prevailing organizational and social conditions happen to be conducive to such usage. But while some have emphasized the potential of the technologies for increased centralization and managerial control, others have suggested that a very different set of outcomes is likely. Davis and Taylor (1975), for example, argue that automated production technologies reduce the need for motor skills and increase the demand for perceptual and decision-making skills. While they recognize the possibilities of alternative applications, for these writers the new technologies are particularly suited to the development of increased employee responsibility and discretion, the creation of semi-autonomous work groups and 'flexible, adaptive, more formless organisations . . . or with bureaucracy based on a consensus and sense of industrial community'. Walton (1982) has made similar observations, arguing that an application of 'social criteria' to the design of new work systems will lead to the development of decentralized, human-centred organizations.

Whatever one's views on what uses *ought* to be made of the new technologies, it is evident that there is considerable variation in the organizational outcomes that information technology may, in fact, bring about. As case reports are beginning to demonstrate (Buchanan and Boddy, 1983) in some instances the technology has been associated with an increased centralization of organizations, in others with an increased decentralization. While deskilling has occurred in some cases, in others more employee discretion has resulted. Attempts to discover a regular pattern of outcomes have proved unsuccessful. What is certain is that, from an organizational point of view, the technologies are extremely flexible in their applications and the organizational outcomes that they may have are not predetermined. There is no simple cause and effect relationship between adoption of the technologies and the structural and performance outcomes that may be associated with them.

Attitudinal outcomes of the technology are also not easy to predict and manipulate. Early reactions to high technology equipment may be dominated by its novelty. Later, a wide range of factors is likely to be important. Included here are the nature of the machinery (how easy it is to use, how intelligible, how functionally adequate), the nature of work practices designed around the technology (including quality of work outcomes and effects on the fabric of the social life of an organization) and the organizational outcomes of the new systems (including factors such as the emphasis that has been placed on different functional priorities in their design and the costs and benefits of the systems to different groups within the organization). The further away one moves from physiological and anthropometric considerations to social and cultural ones the more difficult it becomes to anticipate how others will respond. Similar systems can, in different circumstances, provoke very different responses. What is crucial is how those who are affected by the new systems learn to regard them: whether the new work arrangements are associated with desired outcomes or whether they are found irrelevant or intrusive. Reactions can only satisfactorily be understood within the broader pattern of management and industrial relations of which the new systems form but a part.

Current approaches to the management of IT

It would be wrong to assume that microelectronic systems are always well planned, or that the organizational and performance outcomes that result are always those that were expected. This emerged from a study that the present authors undertook for the Economic and Social Research Council in 1984. Our brief was to study approaches to the evaluation of the introduction of microelectronic technologies within British work organizations, paying particular attention to social and organizational outcomes (Blackler and Brown, 1986). To conduct the survey we identified a number of sectors of the economy where the new technologies are being introduced, interviewing representatives from manufacturing organizations, retailing, telecommunications, central and local government, banking, insurance, health and education, private consultancies and trades unions. It emerged that, while there are differences across sectors, strong emphasis is typically placed on prior-event evaluation with a heavy emphasis on cost-substitution methods. Apart from some technical assessments of the new work systems after their introduction, post-event evaluation, when both 'value added' and the consequences of organizational decisions will be in evidence, is rare. If it does occur, post-event evaluation tends to be concerned with simple cost-substitution calculations only. Beyond some concern with straightforward ergonomic considerations, systematic assessment of the organizational consequences of new technologies is very unusual.

Two general conclusions emerged from the research. First, despite the voluminous literature on computer usage that is directed at management, it would appear that the standard of management of the introduction of new technologies in work organizations is often not high. Writing nearly 20 years ago about management decision-making in general, Lindblom (1959) suggested that decisions consist of a series of disjointed and incremental steps made in an endeavour to achieve marginal improvements. Decision-makers who are operating in such a way have more of an idea of outcomes they wish to avoid than they have about objectives to which they positively aspire. *Ad hoc* tinkering or 'muddling through' is the norm. Lindblom's analysis has been praised by subsequent theorists as an apt description of common practices, although not as a useful model to which managers should adhere.

There is some evidence, however, to suggest that 'muddling through' describes common approaches to the management of information technologies rather well. In addition to the findings of our study of evaluation practices, Viljbrief, Algera and Koopman (1986) have presented a detailed analysis of how decisions were made about the introduction of new technologies within a number of European organizations. They comment on the lack of understanding and control that managers often exert over the process. Similarly, cases reported by Buchanan and Boddy (1983) illustrate how the aspirations of managers towards usage of the new technologies can be somewhat restricted, focusing on short-term considerations rather than long-term ones, departmental needs rather than organization-wide priorities and low-risk rather than high-risk performance outcomes. The research review by Northcott and his co-workers (1985) into acceptance of the new technologies within work organizations within the UK arrived at similar conclusions. They found the extent of take up of the technologies limited at the present time, but most of the applications which have taken place in the UK have been relatively simple ones. To date, only marginal changes to existing organizational arrangements have been achieved with microelectronics-based work systems.

Second, the organizational options made available by the new technologies appear to be poorly understood by many policy makers. While it has come to be recognized that 'muddling through' is no longer an adequate response, management remains fixated on the technology itself and continues to underestimate the psychological and social implications. It is commonly assumed by top management that using the technology to replace people represents the 'state of the art' as far as its usage is concerned. While, as we discuss later, there is reason to suppose that such an approach can be successful in certain situations, in others the weight of evidence suggests that replacement of a 'management control' orientation with an 'employee teamwork approach' can be highly beneficial. Davis and Taylor identified opportunities for organizations to devolve responsibilities and to build consensus, but the majority of managers are only dimly aware of them. Important organizational choices are, it would seem, being made by default.

Although it is clearly essential to try to take into account the potential range of psychological and organizational consequences of new technologies in work organizations (Kling, 1983), in practice this may not be an easy thing to do. In the past 20 years organizational psychology has made important contributions to the theory and practice of organization change, work design and organizational structuring and to approaches to the diffusion of new ideas and techniques about management (see Blackler and Shimmin, 1984, for an introductory overview). In the following sections, insights drawn from the subject are applied to the new technologies. Organizational psychology can, at a general level, alert people to alternative approaches to the utilization of the technologies. It offers concepts and techniques that can be used to help people recognize specific options that might be considered in particular situations. In addition it offers approaches to overcoming the problems of change that can easily become associated with innovative changes within organizations.

RECOGNIZING THE ORGANIZATIONAL OPTIONS

Given the impressive performance characteristics of the technology it is understandable that it comes to be thought of merely as extraordinary gadgetry alone. It is assumed that once certain manpower and training implications have been dealt with, IT can simply be grafted onto a work organization. Organizational psychology encourages a different perspective. Rather than thinking of the technology in terms of gadgetry alone, it encourages a view of technology as a pattern of social relationships. When information technology is thought of not as a 'technological fix', but as an integral part of a social system, questions about work roles and employment relationships, organizational effectiveness, approaches to coordination and behaviour control and alternative approaches to management and organization come clearly to the fore.

Such a perspective indicates that, contrary to the vision of the inevitable logic of an 'autonomous technology' that some analysts have assumed, technologies and the pattern of social relationships that come to be associated with them do not 'just happen'. This point is especially apposite for a technology so versatile as microelectronics. Policy decisions present themselves in at least four areas. Alternatives exist regarding what aspects of the technology should be exploited in any given situation; what and whose goals it should be harnessed to serve; who should participate in the process of technological development; and what assumptions should guide the development of organizational arrangements around the new systems.

Given that alternative approaches to the utilization of the technologies are not immediately apparent to people, the authors propose two models to illustrate the wide range of orientations that are possible. Model 1 is entitled the 'Task and technology approach' and Model 2 the 'Organization and end-user approach'. While Model 1 is based on a 'top down' management control approach, Model 2 draws heavily on an organizational psychology orientation to technological innovation.

For simplicity of presentation on Table 1 the cycle of innovation associated with the introduction of new work systems has been divided into four phases. Phase 1 is the initial recognition that use of the technologies might be advantageous. Phase 2 involves a detailed feasibility review leading to discussions about whether to proceed, recommendations and decisions. Phase 3 is when the accepted solution is further developed or the choice of an 'off-the-shelf' system is made. In phase 4 the system is installed and brought into service.

Managerial interest at the first, tentative phase in the innovation cycle is likely to be motivated by an interest in reduced costs, increased flexibility and improved quality combined with a general desire to keep up to date. Alternative approaches that may guide initial thinking about the relevance of the technologies are shown on the models as either the attitude that people are a costly resource to organizations who ought, if possible, to be replaced or controlled more effectively, or the belief that people are a costly resource whose contributions could be enhanced by appropriate work systems development.

At phase 2, when a detailed exploration and prior justification of possibilities is undertaken, two varying approaches to planning are presented. It is normal to express objectives for technological change through a detailed advanced specification of required end-states. Social planning is less amenable to such treatment. As we suggested earlier, people's reactions to new developments are not easily predicted by outsiders. Moreover their knowledge, skills and attitudes can be expected to develop as a result of experiences gained during a significant programme of workplace reform. An alternative approach to the early formulation of specific organizational requirements is the specification not of the details of end-states required, but of the general principles which it is hoped will guide directions in which changes will be steered. As is shown also in the table, at this stage of the innovation cycle related assumptions are either (i) that the feasibility review process should be managed exclusively by a team of experts, or (ii) the assumption that it is important that potential end-users also should be closely involved at an early stage.

Alternative approaches to phase 3, when general plans are translated into detailed designs or an existing product or system is chosen, are summarized as either (i) the assumption that the new technologies should be used to replace people where possible, or (ii) the assumption that they should be developed to assist people's efforts. Task fragmentation and job simplification on the one hand contrast with job enrichment, autonomous work groups and decentralized organizational structures on the other. Contrasting orientations follow when emphasis is placed on the preoccupations of designers on the one hand or on the concerns of end-users and equipment maintenance staff on the other.

Finally, at phase 4, when the new system is installed and brought on line, the results of the two contrasting approaches become plain. Within a 'task and technology' orientation only minor design modifications are now anticipated and, after a trial period, it is expected that responsibility for the system will pass from the design team to line management. An 'organization and end-user' orientation will, however, adopt a more gradualistic approach, emphasizing the importance of ongoing reviews to the operation of the system via pilot projects and incremental approaches to systems change. Within this approach training and end-user needs are accorded a higher priority than within a 'task and technology' approach.

These models are, inevitably, somewhat general in their presentation and, to fit the detail of particular cases, they may need modifications. However, they are intended to provide yardsticks against which people can assess the assumptions that may guide their own practices and those of others. For example, to call in ergonomists late in the innovation cycle when someone at last recognizes the importance of end-user needs is certainly better than nothing. It may nonetheless indicate only a slight shift towards an 'organization and end-user' orientation. Similarly, the use of consultants to smooth the introduction of a system designed by a remote specialist group, or occasional chats by managers with staff during coffee breaks about possible preferences may give a superficial nod in the direction of an end-user orientation, but has little real impact in the medium or long term.

The question arises of the relative merits of the alternative approaches. Model 2 in the table draws upon a number of approaches developed within organizational psychology. It emphasizes the advantages of evolutionary systems planning, organization and job design issues and the need to give adequate resources to end-users. But the potential advantages of a consultative approach may not

Table 1. Alternative styles of planning and assumptions of planners

	Model 1 *TASK AND TECHNOLOGY CENTRED APPROACH*	*Model 2* *ORGANIZATION AND END-USER CENTRED APPROACH*
Phases	**Guiding assumptions and key actors**	**Guiding assumptions and key actors**
1 INITIAL REVIEW	(*a*) Operating conditions (*b*) People are a costly resource to be reduced if possible (*c*) Key actors: top and senior managers	(*a*) Operating conditions (*b*) People are a costly resource to be more fully utilized (*c*) Key actors: initially from any part of the organization then top management
2 EXPLORATION AND PRIOR JUSTIFICATION	(*a*) Tightly prescribed planning objectives (*b*) Central coordination and control (*c*) Expert driven (*d*) 'Most modern' syndrome (*e*) Key actors: managerial project team including technical and financial experts	(*a*) General policy formulation (*b*) Decentralization, staff involvement (*c*) Concern for end users (*d*) System development potential rather than machine capability (*e*) Key actors: a diverse and representative group, or a consulting project group, or a management plus shadow group. Trade union involvement
3 DESIGN OF SYSTEM	(*a*) Machines over people (*b*) Task fragmentation (*c*) 'Clean design' (*d*) 'Final design' (*e*) Key actors: design engineers and technical consultant	(*a*) People to use machines (*b*) Job enrichment, teams (*c*) Operator and maintenance needs (*d*) Incremental and educative design approach (*e*) Key actors: design engineers, technical consultants, behavioural advisers within consultative procedure
4 IMPLEMENTATION	(*a*) Machine capability (*b*) Only minor modifications expected (*c*) 'Once off' skill training (*d*) Responsibility to line management (*e*) Key actors: as phase 3, also line managers and end-users. Trade union negotiations on conditions	(*a*) User support (*b*) Pilot projects used where possible (*c*) Continuing staff and organization development (*d*) Continuing reviews of operation and needs (*e*) Key actors: as phase 3 also line managers and end-users. Trade union negotiations on training, grading, etc.

always be significant. In certain circumstances Model 1 approaches can work very effectively. A detailed report of such an approach has been provided by McLoughlin, Smith and Dawson (1983) in their description of the introduction of a computerized freight information system within British Rail. As a general rule Model 1 orientations are likely to be most appropriate when a straightforward automation of existing practices is a realistic aim for a new work system. But in other instances Model 2 orientations promise distinctive advantages. Thus, Strassman (1985) pointed to the benefits of introducing advanced office automation systems in ways which encourage staff to become intrinsically involved within them. People learn best how to use advanced office systems by example from others, he suggested; the early involvement of key employees in a new office system is therefore very important. For this to happen, scope must be available for such staff to be able to develop uses of the systems for themselves and to redesign their own jobs around the equipment. A similar series of priorities is discussed by Hayes and Wheelwright (1984) in their commentary on ways in which American manufacturing industry can learn from the best of European and Japanese methods. For organizations to compete effectively through their manufacturing function Hayes and Wheelwright argue that a management control orientation needs to be replaced with an employee teamwork approach in which emphasis is placed on staff learning, problem solving and task involvement.

These accounts and others accord well with the tone of our earlier observations: people's reactions to the new technologies are not predetermined. Model 2 approaches place a premium, not only on organizational priorities, but also on the adequacy of the work environment for those involved within it. The evolving and user-centred orientation it represents is likely to be directly relevant to situations where the involvement of employees in a new work system is given importance, where changes in established working practices are required and where the detailed outcomes of the systems design process are not easy to anticipate. It should be emphasized that many applications of the technology fall into such categories. Obvious examples include, in manufacturing, integrated computer-aided design and manufacturing systems where traditional relations between designers and engineers need to be modified substantially, and in offices, advanced automation systems where it is frequently impossible to specify in advance how usage of the systems will evolve, and where the close involvement of end-users is essential for a system's success.

Table 2. "Model 0" approach: Technology-led 'muddling through'

Phases	Guiding assumptions
1 INITIAL REVIEW	Vague awareness that interesting technologies are becoming available
2 EXPLORATION AND PRIOR JUSTIFICATION	Fascination with the technology Expectations that it can be 'injected' within present organizational arrangements Short-term marginal improvements sought Short-term return on investment orientation
3 DESIGN OF SYSTEM	Technological development understood to be controlled by inherent laws Heavy reliance on systems designers Heavy reliance on the promises of suppliers Machines to enable staff economies
4 IMPLEMENTATION	**'Compensation' approach to the unions** **Unanticipated preoccupation with systems debugging** **Unanticipated organizational problems, staff motivation, demarcation disputes, etc.** **Unanticipated need to undertake staff training and to resource end-users.**

We commented earlier on the fact that the introduction of the new technologies at the present time appears often not to be well managed. To enable comparison with the 'task and technology' and 'organization and end-user' approaches, in Table 2 assumptions that are associated with 'technology led' approaches that set out to 'muddle through' are presented. In such cases phase 1 consists of little more than an awareness that some interesting hardware is now available, phase 2 is dominated by a fascination with the novelty of microelectronics, in phase 3 there is a strong reliance on the promises of the suppliers and of systems designers, and phase 4 is marked by systems debugging, demarcation disputes and a failure to train staff satisfactorily. The lack of a purposive orientation that is a feature of this approach tends to push it towards inconsistency and incoherence. The consequences of this are shown in the table in terms of outcomes that are both unexpected and undesired. Lacking as it does an appreciation of how the technologies might be utilized within long-term business and organizational strategies, this is not a model to be emulated. It may be designated 'Model 0'.

DEVELOPING THE OPTIONS

Important as it is to demonstrate at a general level how different assumptions can lead to different approaches to the management of technological innovation, on its own this is not enough. Much detailed work is needed to enable theoretical options to be translated into specific designs. In this section approaches from organizational psychology to four key issues are introduced.

Analysing the roles and methods of expert advisers

Pressures usually exist within an organization which encourage conventional approaches to the technologies. The traditional orientation of functional experts may combine to encourage managers to look no further than a Model 1 approach. As accountants themselves are beginning to recognize (Sheridan, 1985) conventional accounting techniques may not be well suited to the development of strategic orientations towards information technologies. Thus, production costing techniques which identify direct, indirect, fixed and variable costs are of proven value when applied to work systems dependent on labour efficiency; they are less appropriate when quality, service and speed of response are vital. Conventional return on investment calculations are useful when the objective is to maximize short-term profits; where long-term market share and general competitiveness are crucial, such calculations are less applicable. Similarly, approaches that are frequently followed by systems designers can also encourage a preoccupation with Model 1 orientations. In an effort to make the task of systems development manageable, systems designers traditionally seek to follow a 'phased' approach, first breaking down work systems into smaller and smaller pieces, then endeavouring to recombine the sub-routines into a reconstituted whole (Algera and Koopman, 1984). This approach closely resembles the essentials of job simplification techniques (Braverman, 1974) with the likely outcome of 'active systems and passive individuals'. Systems developers are themselves beginning to recognize the risk of such tendencies in their work (Floyd, 1985). But managers need also to be aware of such tendencies and to encourage the search for alternatives.

Participative approaches to technological development

As shown in Table 1, while Model 1 approaches to the introduction of new technology are likely to be managed in a 'top-down' style, Model

2 orientations seek to involve a broader range of relevant personnel. Such an approach has a number of potential advantages, most significantly perhaps in the opportunities it offers for exploring a wider range of design alternatives than might otherwise be considered. Participative approaches can take a number of forms. Mumford (1983) has argued for the benefits of the direct participation of end-users in the design process. Other alternatives include the incorporation of end-user representatives into the design team, procedures whereby design teams consult separately with groups of employees affected by a new system, the designation of individuals within the team to represent the interests of relevant groups who may not themselves be 'computer literate', and an arrangement whereby management and employee groups work separately on their preferred approaches, coming together to negotiate a solution that would be acceptable to both.

Different approaches to participation are likely to be appropriate in different situations. Factors relevant to the choice of alternative approaches include the importance of the project to an organization, whether problems of a technical or applications nature predominate and the chances that significant conflicts of interest will emerge (Opperman, 1986). Clearly, participative approaches to systems development are no panacea, demanding as they do a high level of managerial competency.

Psychological analysis of participative approaches can indicate ways of improving its effectiveness as a technique for extending the range of options a design team can consider. Mambray, Opperman and Tepper (1986), for example, report the findings of research funded by the West German government into the effectiveness of participative approaches to systems development. They report how, in the cases that they studied, end-users tended not to have a very clear idea of the technical options that they could consider nor of the range of information technology products currently available. They tended to be considerably influenced by the very positive presentations of developments offered by management and to assume a short-term orientation to the changes. Senior managers, though privately acknowledging that the new systems represented the start of comprehensive new approaches to organization, tended in their public utterances to reinforce such a 'minimalization' perspective. And in line with the points made earlier in this chapter, Mambray reports how both managers and end-users assumed that the process of systems development was controlled by inherent rules and tended, by the deference they showed to systems engineers, to reinforce them in their expert roles. In the context of the history of approaches to participative management, such findings are, perhaps, unremarkable. Nonetheless, they indicate that if participation is to be effective as a technique for generating design possibilities, it requires training, support and resources.

Approaches to evaluation

Earlier we noted how, in the UK at the present time, managerial approaches to the evaluation of new technologies tend to be restricted to short-term cost substitution techniques. The themes of this chapter suggest, however, that a more central role for evaluation would be beneficial. Qualitative approaches to evaluation can focus attention on both the 'value added' and the behavioural consequences of new work systems. Moreover, the themes of this chapter illustrate how the standard of the management of information technologies is related to the general standard of management practices in this country.

The task of introducing new work systems itself offers considerable opportunities for individual and organizational learning, and evaluation can play an important role when used as part of a cyclical approach to systems development. Psychological approaches to the evaluation of the new technologies have already been developed. Thus, Damodoran (1981) presents an approach that concentrates on the technical adequacy of new equipment, its ease of use, the quality of support provided for end-users and job design outcomes. More detailed evaluative schemes may be of value within particular situations. But what does seem crucial is that management should begin to accept the need to review systematically the reactions of their staff and to treat the data so produced with some priority.

Encouraging adoption of new approaches

In considering how interest in new ideas about the management of information technologies can be encouraged it is tempting to think that publication of case examples of their success is likely to be effective. Undoubtedly case reports can have an impact on public opinion. But experience with approaches that depend heavily on such a strategy for the diffusion of new practices suggests that they are limited in their effectiveness. It seems (Herbst, 1974) that people often feel that the cases reported differ in important particulars from their own situations or are special in some way. Moreover, success is often attributed to the talent of the consultant expert involved in the case rather than to the general validity of the ideas guiding his or her practice. Typically, people do not seem to identify easily with examples of social and organizational change that are presented by experts as examples that others should seek to emulate.

As an alternative the development of networks can be encouraged to keep people in touch with useful developments that may be of interest to themselves. 'Workshops' can be arranged where managers, employees and unionists who have had experience with new approaches for the introduction of new technologies can discuss them with others who are considering embarking on similar ventures. Such an approach can avoid the impression that such new approaches are the exclusive property of outside and remote 'experts' and can do much to encourage a rapid diffusion of new ideas. 'Package' solutions that consultants may propose are less likely to appear attractive when such an approach is adopted. Having embarked on a 'search' process that they are controlling themselves, people are likely to be both more creative and assertive in the development of solutions that suit their own particular needs.

INTRODUCING INNOVATIVE APPROACHES

Even though general alternatives towards utilization of the technologies may have been recognized and detailed work has translated these into design options, problems can still emerge. It is a mistake to assume that the advisability of well-thought-out changes will inevitably impress others. When the technologies are thought of not as mere gadgetry but rather as patterns of social relationships it becomes clear that the dynamics of organizational life can be expected to inhibit enthusiasm for certain types of proposal. Changes in established roles, relationships, patterns of communication, influence and control are to be expected from the new technologies. Not all changes will be resisted of course; but changes that promise to threaten key purposes and attachments of those affected by them will be resisted. Pressures to maintain the status quo in organizational life can be such that important opportunities for organizational reform may be lost.

In considering how to approach the 'dynamic conservatism' (Schon, 1970) of organizations it is useful to recognize that organizations are not unitary structures with one focus of loyalty for their members. Rather, they are composed of individuals and groups with differing priorities and interests, who have varying opportunities to secure their aspirations. Differences of interest within organizations are rarely settled by a naked exercise of power, of course; in most instances they are resolved by the exercise of authority or by processes of 'organizational politics'. Typically, political activity occurs in organizations in situations in which the addition of further information alone will not solve a particular problem (Pfeffer, 1981). Examples of organizational politics include, at the individual level, efforts of junior staff to impress their seniors and to develop links with them and, at the group level, efforts to co-opt dissenters or to discredit them, the development of temporary coalitions, and the selective use of outside experts to back the objectives of one faction or another.

Keen (1981) discusses efforts to introduce management information systems that turned out to be 'technical successes and organisational failures'. He comments that it is unfortunate that organizational politics has been 'equated with evil, corruption and, worst of all, blasphemy in the presence of the Rational Ideal . . . politics are the process of getting commitment, or building support, or creating a momentum for change: they are inevitable'. Taking this view it is foolish to rely either on the power of reason to justify suggestions for organizational reform, or on the authority of management decisions. Rather, the inevitability that different views will be held on the desirability of significant organizational changes should be recognized, as should the inevitability that people will seek to protect themselves from future disadvantage. Kling (1985), indeed, suggests that it is useful to draw an analogy between the management of large systems development projects and the management of urban development programmes. The designers of both, he suggests, are producing 'multi-faceted artefacts for plural, conflicting, and not wholly determined uses where the inhabitants are not the designer's "customer" and where tradeoffs are inexact'. Urban planners have in recent years, however, become familiar with the need to consult with those affected by their work; Kling's proposal is that systems designers should adopt similar approaches.

Regarding the phenomenon of 'resistance to change' Keen (1981) describes how effective the 'tactics of counter implementation' can be. Thus, a project can be lost should it be made to appear complex, controversial, hard to coordinate or vague. The legitimacy of technical experts can be called into question, and the lack of inside knowledge that outsiders display can be exploited. Rather than regarding counter-implementation tactics as entirely aberrant, however, approaches from organizational psychology seek to encourage a direct articulation of grievances or uncertainties and to develop alternative solutions. In place of covert attempts to influence the course of events, a pluralist orientation can create an environment where differences of interest can be recognized, articulated and resolved. Wherever possible, such approaches encourage an inclusive orientation to the management of change, in which the priorities for improvements are decided by the people who are directly affected. From this starting point the task of mobilizing effort, identifying feasible objectives and integrating change into established practices can be relatively straightforward (Blackler and Shimmin, 1984). Where change needs to be initiated from the top, then a psychological orientation indicates the advantages of early discussion with those likely to be affected, leading to an agreement that something has to be done, co-option of those affected early in the planning process, and the visibility, acceptability and accessibility of managers in charge of the project (Keen, 1981).

The term 'resistance to change' has often been used in the management literature to refer to the likely reactions of junior members of an organization to innovations desired by senior personnel. It may be noted, however, that a number of the more important potential uses of the new technologies require fundamental changes also in the roles and behaviours of senior management staff. Various commentators have observed that adoption of the new technologies in Britain may be inhibited by traditional values. Job specialization has its roots deep in the structure of established roles within British work organizations and is perpetuated by education and training practices, as well as by divisions in the society. Reinecke's (1984) description of the conservatism amongst office staff and the low status typically associated with keyboard skills (essential to effective usage of advanced office equipment) is a case in point. In manufacturing, Senker's (1984) analysis of certain advanced technologies indicates how they may require higher levels of technical knowledge at senior levels of an organization than are normally found in British firms. Commenting on the general significance of traditional organizational practices for effective usage of new technology, Dore (1983) has suggested that the apparent reluctance of the British to move from traditional forms of industrial organization is having an increasingly deleterious effect on the ability of British companies to compete in international markets.

It would be wrong therefore to conclude that adoption of psychological approaches to organizational change merely helps ensure commitment to decisions already reached by management. A recognition of the indeterminism of the technologies and an approach to their introduction which encourages the detailed exploration of possible options fosters an environment in which important questions may be addressed. The approach suggested in this chapter can, on the one hand, encourage a sceptical approach to the extravagant claims that are sometimes made about microelectronics and, on the other, encourage people to question the relevance of approaches to work and organizing that were developed prior to the appearance of information technologies.

SUMMARY

In this chapter it is suggested that:

1. The technical capabilities of microelectronics are such that they offer considerable opportunities to improve the performance of work organizations. While short-term cost reductions can be important, 'value added' and organizational benefits are also likely to be highly significant for advanced applications of the technologies.

2. It is a mistake to assume that the mere purchase and commission of the new technologies will automatically enhance an organization's effectiveness. Sharply contrasting views have been proposed about the form organizational changes associated with the technologies are likely to take. The technologies can be used to centralize organizations and to increase managerial control or to improve organizational integration and to decentralize decision-making. Key policy options present themselves for the utilization of the new technologies but very significant choices are presently being made by default.

3. It appears that present approaches to the design and introduction of the technologies are often not of a high standard. A 'technology-led' style of 'muddling through' appears common.

4. Alternative approaches to the technologies are illustrated by two models. Model 1, the 'task and technology-centred approach', emphasizes the advantages of tighter management control, specific planning objectives, design priorities and 'once off' systems development. This orientation is perhaps the most clearly recognized alternative to 'muddling through'. Model 2, the 'organization and end-user approach', is developed from theory within organizational psychology. It emphasizes the advantages of organizational integration,

general planning objectives, end-user needs and an ongoing emphasis on systems development.

5. Much detailed work is needed to explore the design options that are feasible within any particular situation. Approaches from organizational psychology that can be of assistance include analysis of the roles and methods of key functional advisers, the development of approaches to design which involve end-users and the development of evaluation techniques.

6. Resistance to change can occur when established priorities and methods are threatened by proposals for change. Approaches to organization which were developed before information technology may need to be discarded. A pluralist orientation to technological development is required to make this possible. The long-term issue is not 'how fast can British work organizations automate?' The issue is 'how can they automate well?'

REFERENCES

Advisory Council For Applied Research and Development (ACARD) (1979) *Technological Change: Threats and Opportunities for the United Kingdom*. London: HMSO.

Algera, J.A. and Koopman, P.L. (1984) Automation: Design process and implementation. In P.J.D. Drenth, H. Thierry, P.J. Willems and C.J. de Wolff (eds) *Handbook of Work and Organisational Psychology*. Chichester: Wiley.

Blackler, F. and Brown, C.A. (1986) Current British practices in the evaluation of the new information technologies. In H.W. Schroiff and G. Debus (eds) *Proceedings of the West European Conference on the Psychology of Work and Organisation*. Amsterdam: North Holland.

Blackler, F. and Shimmin, S. (1984) *Applying Psychology in Organisations*. London: Methuen.

Braverman, H. (1974) *Labour and Monopoly Capital*. New York: Monthly Review Press.

Buchanan, D.A. and Boddy, D. (1983) *Organisations in the Computer Age: Technological Imperatives and Strategic Choice*. Aldershot: Gower.

Child, J. (1984) New technology and developments in management organisation. *Omega: the International Journal of Management Science*, 12, 211–223.

Cooley, M. (1981) *Architect or Bee? The Human/Technology Relationship*. Slough: Langley Technological Services.

Damodoran, L. (1981) Measures of user acceptability. In B.G. Pearce (ed.) *Health Hazards of VDUs*. Chichester: Wiley.

Davis, L.E. and Taylor, J.C. (1975) Technology effects on job, work and organisation structure: a contingency view. In L.E. Davis and A.B. Cherns (eds) *The Quality of Working Life, Volume 1*. New York: The Free Press.

Dore, R.P. (1983) *The Social Sources of the Will to Innovate*. London: The Technical Change Centre.

Floyd, C. (1985) Towards a paradigm change in software engineering. Paper presented at the conference on Development and Use of Computer-Based Systems and Tools, Aarhus University. Available from the Department of Computing, Aarhus University.

Hayes, R. and Wheelwright, S. (1984) *Restoring our Competitive Edge*. New York: Wiley.

Herbst, P.G. (1974) *Alternatives to Hierarchies*. Lieden: Martinus Nijhof.

Keen, P.G.W. (1981) Information systems and organisational design. *Communications of the Association for Computer Machinery, 24(1)*, 24–33.

Kling, R. (1985) Computerisation as an ongoing social and political process. Paper presented at the conference on Development and Use of Computer-Based Systems and Tools, Aarhus University. Available from Department of Computing, Aarhus University.

Leavitt, H.R. and Whisler, T.L. (1958) Management in the 1980s. *Harvard Business Review, 36*, 41–48.

Lindblom, C.E. (1959) The science of muddling through. *Public Administration Review, 19*, 79–88.

Mambray, P., Opperman, R. and Tepper, A. (1986) Experiences in participative design. Mimeo available from Gesellschaft für Mathematik und Datenverarbeitung, Bonn.

McLoughlin, I.P., Smith, J.H. and Dawson, P.M.B. (1983) The introduction of a computerised freight information system in British rail. Mimeo available from Southampton University New Technology Research Group.

Mumford, E. (1983) Participative systems design: Practice and theory. *Journal of Occupational Behaviour, 4*, 47–57.

Northcott, J., Forgarty, M. and Trevor, M. (1985) *Chips and Jobs: Acceptance of New Technology at Work*. London: Policy Studies Institute.

Opperman, R. (1986) User participation: Some experiences and recommendations. Mimeo available from Gesellschaft für Mathematik und Datenverarbeitung, Bonn.

Pfeffer, J. (1981) *Power in Organisations*. Marshfield, Mass: Pitman.

Reinecke, I. (1984) *Electronic Illusions*. Harmondsworth: Penguin.

Schon, D. (1970) *Beyond the Stable State*. Harmondsworth: Penguin.

Senker, P. (1984) The implications of CAD/CAM for management. *Omega: The International Journal of Management Science*, 12(3), 225–232.

Sheridan, T.J. (1985) Financial management and innovation – a need for change. *Proceedings of the Royal Aeronautical Society Spring Convention*. London: The Royal Aeronautical Society.

Strassman, P. (1985) *The Information Payoff*. New York: Free Press.

Viljbrief, H.P.J., Algera, J.A. and Koopman, P.L. (1986) The management of automation projects. In H.W. Scroiff and G. Debuss (eds) *Proceedings of the West European Conference on the Psychology of Work and Organisation*. Amsterdam: North Holland.

Walton, R.E. (1982) New perspectives on the world of work: Social choice in the development of advanced information technology. *Human Relations, 35(12)*, 1073–1084.

MANAGING FACTORY AUTOMATION

Christopher Clegg and Toby D. Wall

This chapter examines the management side of new factory automation systems. Running throughout is the concern that management issues, in contrast to technical ones, have not received the attention they deserve, either from those with a practical responsibility for the effectiveness of factory automation systems, or from those with a research interest in their development and use. The aim is to identify some of these management issues, and show how they may be addressed.

The chapter is in four parts. The first sets the scene by describing the main forms of new factory automation, outlining the extent of their penetration into British industry. The second argues that research, development and practice in this area have been dominated by technical concerns to the unfortunate exclusion of debate about managerial and organizational factors. This is followed by identification of the key management issues of importance both for the overall performance of such systems and for the mental health of those operating them. This part introduces the notion of choice in how to manage and organize factory automation systems, and considers the effectiveness of two main alternatives. The fourth part considers the future of such systems, and the chapter concludes by arguing for a more effective and psychologically healthier approach to their management than typically prevails.

Throughout the chapter the term 'management' is used in its broadest sense. It covers all aspects of the design of the organization and of people's jobs within it, from selection and training, through the specification of operator, supervisory and managerial roles, to the design of appropriate organizational structures. The argument is for a particular approach to the management of new factory automation for which there is growing support, that is one which promotes operator control. Nevertheless, given that this technology is still in the early stages of development, these views are not established 'facts'. Our purpose is thus to raise issues and options which might otherwise be neglected.

FORMS AND DIFFUSION OF FACTORY AUTOMATION SYSTEMS (FAS)

The term 'factory automation' is used to refer to advanced manufacturing systems based on information technology. It is recognized that other forms of automation exist, but the focus here is on computer-aided systems directly involved in the manufacturing process. Whilst this description concentrates on the most common types of such automation, the arguments also apply to less prevalent forms, as well as to tailor-made systems developed within individual companies to meet unique and particular needs.

A large-scale industrial survey conducted in 1981 by Northcott and colleagues at the Policy Studies Institute (see Northcott and Rogers, 1984; Northcott, Rogers and Zellinger, 1984) provides a general context for this chapter. This survey covered a sample of 1,200 factories representative of the full range of British industry, in order to determine the pattern of use and impact of microelectronic applications in manufacturing processes. It was found that approximately 40 per cent of British factories use microelectronics in manufacturing, though users do not necessarily do so on a large scale. In practice, microelectronics is used in around 18 per cent of all production processes. Not surprisingly, very large factories are much heavier users than small, and applications are widespread across industry types. Nevertheless there are differences across sectors; for example, engineering, food and paper, printing and publishing are relatively heavy users.

Looking at specific functions, the most common usage is for control of individual machines or processes (in use in three-quarters of factories using microelectronics), whilst only one in six users applies them to more ambitious integrated control of groups of machines or of several stages of production. Turning to specific types of technology, the most widespread are computerized numerically controlled machine tools, of which there were around 17,000 in 1983, predicted to increase to 27,000 by 1985. During the same period, computer-aided design applications were predicted to increase from 8,800 to 14,000, and the 'symbol' of high technology, the industrial robot, from 900 to 4,700.

Given the variety of applications and predictions concerning future developments in factory automation, it is useful to view these emergent technologies from an evolutionary perspective. In this light, three major generations of computer-based FAS can be identified. The first is characterized by stand-alone computer-aided equipment. This is the stage to which most companies with experience of FAS have now evolved. As has been noted, its most common form throughout the world, including the UK, is the Computerized Numerically Controlled (CNC) machine tool for cutting and shaping metal. A CNC machine is like a traditional manual machine in its preparatory setting up activities, but thereafter it is controlled by a computer program rather than manually, and thus requires programming. Furthermore, since such programs often require correction or improvement, they need validating (or proving out) and then editing. CNC machines offer benefits in terms of excellent consistency, repeatability and flexibility. At this stage of development, what typically happens is that CNC machines are purchased as replacements for and improvements on existing manual machines. The diffusion of such technology is accelerating rapidly.

The second identifiable generation of computer-based factory automation sees the integration of several stand-alone pieces of equipment into an overall operating system. The most advanced form of such systems is the Flexible Manufacturing System (FMS). An FMS typically incorporates several stand-alone pieces of computerized technology, such as CNCs and robots, integrated by a computer system and by the provision of some automatic transfer capability between its different parts (for example automatic guided vehicles, conveyor belts). The computer enables programming of individual components, but above all allows the control of work around the system. However, whilst there has been much interest in and publicity about the integrated FMS, there has been less action within companies than one might have been led to believe. Indeed, Bessant and Haywood (1985) estimate that worldwide there were only around 200 operational systems in 1984, of which approximately 10 existed in the UK. Whilst FMSs are notoriously difficult to define (it is one of their most flexible features), it is apparent that most manufacturing industry has not evolved to this stage.

The third identifiable generation of new FAS goes one step further with regard to system integration. In this case several activities traditionally seen as indirect to, and conducted separately from, the production process are directly integrated into it. For example, raw materials and stores management procedures are included in Computer Integrated Manufacturing (CIM) systems to allow for the automatic reordering of materials, stock control and the management of inventory. Similarly automatic diagnostic and correction aids may be incorporated to provide for improved maintenance. It is this level of integration that has led to speculation about the 'people-less' factory, but at present such levels of automation are at the research and development stage. Inevitably perhaps, CIM systems attract considerable publicity. However, it is anticipated that whilst a few prototypes will be developed and built over the next 10 years, the rates of diffusion will be quite slow due to the costs involved and the sheer complexity of the design problems.

Whilst these comments may sound rather conservative or pessimistic, depending on one's perspective, they are not intended as such. Many UK organizations are actively considering, developing and experimenting with some form of FAS, and typically are running them in parallel with their tried and tested 'old technologies'. In medium-sized and large organizations in particular, an incremental approach to FAS seems likely to characterize the way forward. Thus many companies are developing islands of automation alongside older technologies. For those in the forefront of technical change, it is likely that their systems will become larger, more integrated and more complex. It may also be predicted, partly because of the size and importance of the market, that such technologies will increasingly be applied to small batch production environments. These issues are considered later in the chapter, but here it should be noted that this is probably an ideal time for identifying and debating the management issues and problems that have been experienced by those companies who, relatively speaking, are in the vanguard of technical innovation.

'TECHNICAL MYOPIA' IN MANAGING CHANGE

The point here is straightforward: most companies developing and implementing FAS see the issues and problems as essentially technical in nature. The motives for adoption may vary from keeping up with competitors, through trying to improve quality and productivity, but the emphasis and effort is very much technically based and technically driven. The management aspects of such technologies are ignored. An example from the authors' own research and development experience illustrates this argument (see Clegg and Kemp, 1986).

One 'high-tech' engineering company was moving rapidly from manual production processes, through stand-alone CNC technologies, towards an integrated FMS. The process lasted several years and was, in its formative stages, designed, controlled and coordinated by a project team comprising a range of technical specialists. In practice, the design was driven and dominated by these specialists, to the exclusion of those with other experiences and perspectives, such as line managers, personnel and industrial relations experts, and user representatives. Such exclusions occurred because their inputs were not seen as particularly relevant, and the engineers believed their involvement would be inefficient, slowing down the design process because of lack of knowledge and relevant technical expertise. As a result the broader human aspects of such systems were hardly considered in this design process, and there was little explicit consideration of how the systems would be operated and managed, an issue described more fully in the next part of this chapter.

Of course not all human aspects were ignored. In the early stages discussions covered the numbers of employees required to operate the system, since this was a major financial aspect of the investment. Later on there was also explicit consideration of the training required for the people who would run the system. But there was no debate about how the system should be managed and operated until the technical designs were finalized. This can be described as the 'sequential approach' to system design and is summarized in Figure 1. It is the most common approach to the management of technical change in manufacturing companies, as several commentators have observed (see Blackler and Brown, 1986).

The next question is whether it matters that design and implementation happen in this way. It does, for a number of reasons.

First, if there is no overall plan for the managerial aspects of the new technologies, people in different parts of the organization may work with incompatible assumptions about the goals they are pursuing. An example illustrates the point. In one company moving towards a high level of factory automation, some people were aiming for the new technology to deskill operators, whilst others in the same

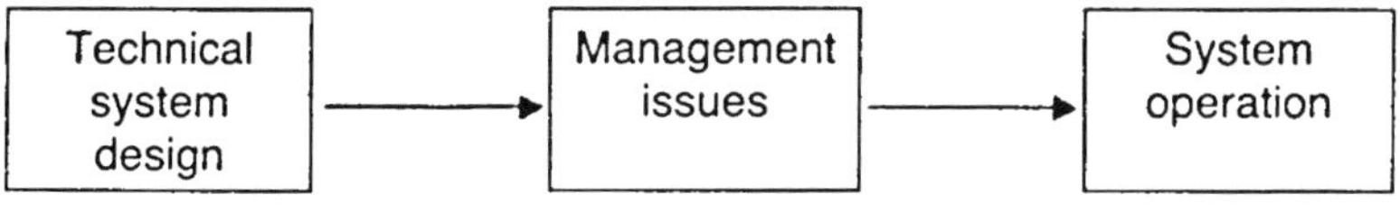

Figure 1. Sequential approach to designing FAS.

factory were working towards the opposite goal of enhanced skill levels (see Mueller *et al.*, 1986). This led to difficulties during the planning stage, and problems once the system became operational. In large, busy production environments, it is important to have an overall management plan around which the various contributors to change can integrate their efforts. If not, the process becomes highly fragmented and may well be sub-optimal.

Second, this approach can mean that the discussions of the managerial aspects are delayed until the last moment and 'squeezed' into unreasonably short periods of time. This is often the case with training. Indeed it is not uncommon to hear people complain that most of the managerial issues are not discussed until after the system is in operation!

Third, such absence of managerial planning may lead to problems of demarcation once the systems become operational, simply because the issues have not been aired and discussed much earlier in the process. Certainly there is evidence that the effectiveness of new technology has been reduced because, once commissioned, different groups of employees have been in competition for undertaking various activities concerning its operation and control. Such issues include who should have responsibility for programming, editing, and routine maintenance of the systems. This problem may be compounded when several trade unions are involved.

Fourth, there is evidence that preferred methods of operating and managing systems can be pre-empted by the design of the system itself, in other words because these managerial issues are not addressed early enough in the design process (Corbett, 1986; Rosenbrock, 1983).

And finally, such an approach may simply be sub-optimal because poor design choices are made in the absence of people with the expertise to discuss how such systems should be managed. As any production manager knows, the performance of technology is not defined on the 'drawing-board', and the practical realities of the manufacturing process need to be taken into account.

Overall it is clear that many new systems are designed with too little regard for their management and operation, and such consideration as these issues receive is undertaken too late. At least 90 to 95 per cent of company resources (time, energy, expertise and money) are spent on the technical aspects to the exclusion of these managerial issues. With the limitations of this strategy in mind, a 'parallel design approach' is advocated (see Figure 2), wherein the managerial and technical aspects are considered at the same time, either by the same

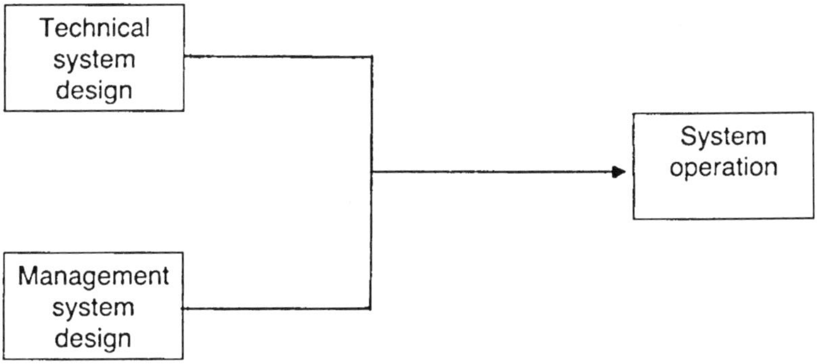

Figure 2. Parallel approach to designing FAS.

groups of people or by different groups with planned means of coordinating their work (Clegg and Kemp, 1986).

This approach reflects the logic of the 'socio-technical' perspective, popular in the organizational literature more generally (see, for example, Taylor, 1975; Cherns, 1976; Trist, 1978). This holds that new systems are not simply technical entities, and that they can only be successful in their operation if *both* the technical and the social (that is managerial/human) aspects are considered with the aim of their joint optimization.

MANAGING FACTORY AUTOMATION SYSTEMS

The sequential approach to technical change means that most companies do not explicitly consider the optimum way to manage and operate their FAS. Some simply use the same methods that they used with their older technologies, and many 'drift' into a management approach. In practice, of course, there can be a broad variety of ways to manage such systems. Nevertheless, it is now clear that, in practice, these fall into two main types (for example, Dodgson, 1985).

The alternative approaches to the management of FAS can be illustrated by description of practices within two companies representing each of these two types. These may be labelled 'Specialist Control' (as practised in company 1) and 'Operator Control' (as practised in company 2). Both companies use CNC equipment and they provide an ideal focus in this context, because CNCs are the most prevalent form of FAS and represent the technology about which most is known. Both enterprises operate in small batch, high precision engineering markets, making products for the aerospace industry. The approaches are described under six headings.

1. *Philosophy and objectives*

In company 1 (adopting the Specialist Control approach), the logic management use is that CNC machine tools are an expensive investment which therefore needs to be controlled directly by experts rather than by shopfloor operators. The overall philosophy is that investment should be in the expertise of people in specialist departments who can control the operation of the technology. Typical objectives in this approach are to maximize the level of management control and to achieve high performance through specialization.

In company 2 (representing the Operator Control approach), the logic is that CNC machines are certainly expensive but therefore need to be manned and controlled by expert operators who can respond to problems as they occur. The philosophy is that operators are local experts in their own right, needing the support of people in the indirect functions only for more fundamental problems. Objectives in this approach are to achieve quick responses to problems by getting operators to control 'variances at source' and to foster employee motivation and commitment to quality.

2. *Method and structure*

In company 1, the method of pursuing the objectives is to specialize using a differentiated organizational structure. For example, there are separate groups of tool-setters and machine-setters, as well as a large department of programmers. The setters do all the tool and machine setting, and the programmers have responsibility for preparing all the CNC tapes, as well as for validating and editing them. For their part the operators have no access to the tapes, which are locked into the machines by the programmers, and they work as machine minders calling on external experts when problems arise. This represents the classic 'deskilling' approach to the operation and management of new technology based on the 'division of labour' (see Braverman, 1974). The benefits and costs of this managerial style are considered below (section 4), but the major problems to highlight here concern the difficulties of coordinating the efforts of the different specialist groups, and the psychological impact of the deskilled jobs on the motivation and overall mental health of the operators.

In company 2, the objectives are pursued using a much simpler organizational structure. There are no specialist machine-setters, only one tool-setter (who makes tools for specialist jobs), and a relatively small department of programmers. The programmers prepare the tapes, but the operators are responsible for their validation on the machines and for much of their subsequent editing. In addition, operators write simple programs when the need arises, and undertake all their own machine setting and most of their own tool setting. In this company the operators are flexible: they see themselves and are treated as relatively skilled engineers. The major managerial problems with this approach concern the availability of the necessary skills and the difficulty of maintaining control of the production process (see section 4).

One way of capturing the important differences between these two companies is to consider the problem of tool wear. In engineering work of this kind, cutting tools inevitably wear down with repeated machining and this affects cutting dimensions. In company 1 this problem is seen as one for the specialists to resolve and the programmers write three different programs for each job depending on the level of wear on the tool (light, medium or heavy). The program is selected according to the wear of the tool. In company 2 the same problem is treated quite differently. Here the operators make the necessary adjustments to the machines to allow for the wear of the tool in use (see Kemp and Clegg, 1986).

3. *Personnel plans*

Organizations choosing the Specialist Control approach adopted by company 1 have to make various detailed plans and the focus here is on those relevant to the personnel function. Training for operators is typically only limited and focuses on the level of vigilance required to identify machining problems as early as possible in the cutting process. It may also involve recognition of different sorts of problems so that appropriate experts can be called in. Usually operators are relatively unskilled and are graded and paid accordingly. Much more training is required for specialists in the other departments, and there may be difficulties in identifying the comparative grading and payment between these different groups. For the personnel department, careful planning is also required for any industrial relations aspects of this differentiated structure. For example, it may be that several different unions are involved, and lengthy consultations and negotiations may be required to identify and agree upon the distribution of responsibilities.

For organizations choosing the Operator Control approach of company 2, the detailed personnel plans are different, although in the same areas. Extensive training is required for the skilled operators and it is necessary for their skills to be updated periodically. The job gradings and payment need to reflect the level of skills, and high payment may be one method that managers use to prevent their most skilled employees moving on to the other companies where skills shortages may exist. Such comparability problems as do occur within the company are likely to be between the most flexible and expert operators and those with less expertise.

4. *Perceived benefits and costs*

In company 1, managers believe a number of benefits accrue from their choice of Specialist Control. The direct labour costs of unskilled operators are relatively low, as are the costs of recruiting and training newcomers. Managers feel they have 'direct control' of the production process, or at least that it is in the hands of specialists. This has the perceived benefit that operating disciplines are relatively tightly maintained and that responsibilities can be clearly defined.

On the other hand there are costs or disadvantages with this approach. The indirect costs are relatively high, largely because there are large numbers of indirect staff. Furthermore, the operators may be poorly motivated by the deskilled nature of their jobs. Problems can also occur in coordinating the work of the separate indirect functions, which typically have quite different sets of objectives and priorities, as well as different views of difficulties on the shopfloor. Such a method of organization can also lead to relatively poor machine utilization, because there is always waiting time before machine problems are rectified, and the operator, in ignorance of the cause of an operational problem, may call the wrong specialist.

In company 2, the profile of benefits and costs is quite different. Typical reported benefits of the Operator Control approach include relatively low indirect costs because of the low indirect staffing levels. In addition, high levels of motivation and commitment to quality exist on the part of the operators. A representative quotation from one of the CNC machinists in answer to the question 'What are the best parts of your job?' gives a flavour of their views.

> The best parts of my job are when I get interesting one-off components to manufacture, as I seem to get very bored with mass production work, but give me an intricate and challenging

one-off job and I get so much satisfaction out of doing that job than anything (*sic*) (Kemp and Clegg, 1986).

The other major benefit of this approach can be found in the high levels of machine utilization which result from the quick resolution of production problems by the operators.

Again, however, this method is not without its disadvantages. From a managerial perspective, the direct costs of skilled labour are relatively high, as are the costs of recruiting and training newcomers. Management also feel that they have less 'direct control' of the production process since so much is controlled on the shopfloor. They also worry that operators take 'shortcuts' in some of the operating disciplines that time-served specialists would not.

5. *Choice and Circumstances*

It is apparent then that companies 1 and 2 have made different choices for how they manage and operate their CNC machine tools. A key question is, why?

Company 1 is a large corporation with a bureaucratic structure and orientation, and with a strong and differentiated pattern of trade unionism. On the other hand, company 2 is a small company, albeit a part of a large multi-national group, but its local history and culture are more paternalistic and its structure more organic. It has no history of trade unionism. In both cases the companies manage and operate their new technologies as they do their old.

More generally, it is possible to identify three interrelated sets of factors which influence the choices companies make when implementing FAS. The first concerns characteristics of the company. Important factors here include the size of the firm and the factory in question, the existing ways of managing the production process (that is custom and practice), the level of bureaucracy and specialization in the factory, the local culture and ideologies and the pattern of trade unionism in both the company and the local factory. The second set of factors concerns the characteristics of the product and the market. Important features include the batch sizes that are manufactured and the complexity and quality demanded of the product by the market. The third set of influences involves the circumstances outside the company, a key feature being the availability of skilled labour (see Child, 1984; Sorge *et al.*, 1983).

To be more specific, companies are more likely to manage their FAS using the Specialist Control approach when:

— this reflects their existing way of managing the production process
— they are relatively large and bureaucratic
— this method matches the local culture
— there are several strong trade unions
— the batch sizes are large
— the product is relatively simple to produce and the quality standards not all that exacting
— there is a shortage of skilled labour.

The Operator Control approach typically occurs in the opposite circumstances.

Of course these factors may combine in much more complex ways than the above account suggests. For example, it may be that a firm is small, bureaucratic, has several strong trade unions and makes very complex products in small batches. From what we have said it is not clear what predictions one may make regarding the likely choice management would make to manage and operate their FAS. These multiple influences merit further consideration. However, this historical perspective on why particular approaches emerge, though interesting, is not the most important issue. More salient here is to determine which approach is more effective, and under what circumstances, both with regard to the performance of the system and to the mental health of the operators.

6. *Performance and mental health*

Looking first at the performance of FAS, a distinction should be drawn between two sorts of production environment in terms of their levels of 'uncertainty'. High levels of uncertainty exist when batch sizes are small, products are very complex and difficult to produce, quality standards are exacting and when the production process requires the input and collaboration of different skills and expertise. Relatively low levels of uncertainty exist in the opposite circumstances, (big batches, simple products, low quality standards, etc.). High levels of uncertainty simply mean that the production process is more difficult to manage because it is highly unpredictable (see Galbraith, 1977).

In relatively certain/predictable production processes, the two choices are potentially equally effective. The Specialist Control approach works effectively because management and the various specialists can predict what problems will occur and can implement necessary procedures and systems to handle them. The Operator

Control approach works because the operators can resolve the difficulties themselves.

However, in relatively uncertain production processes, according to the criteria specified above, the Operator Control approach is likely to be the more effective. This results from the ability of skilled operators to resolve problems as they occur, and from the motivational benefits which accrue from this method of organization. Thus operators with the requisite skills and responsibilities can attend immediately to the (more frequent) adjustments to products, programs and technology in order to minimize downtime and maintain quality. Moreover, because of the motivational benefits of the more highly skilled work, they are more attentive and creative in preempting production problems. No matter how expert the specialists, they cannot devote the same energy and attention to everyday problems, which is why the Specialist Control approach is necessarily less effective in this environment. Thus performance benefits from the Operator Control approach in uncertain environments arise from better system utilization and improved product quality (see Jones, 1984). These observations are summarized in Figure 3. Both the lines on the graph slope downwards because performance will be worse in uncertain environments than in predictable ones.

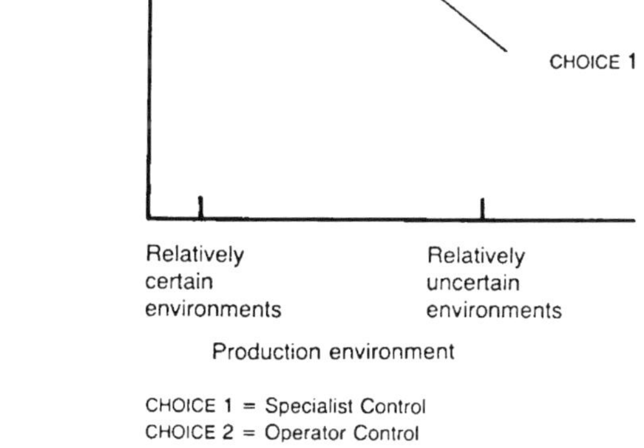

Figure 3. Choice and performance.

Turning to the mental health of people working on FAS, it should be noted that this term is used here in a general sense. It includes attitudes to work (such as job satisfaction), motivation at work and various affective reactions to the job. A relatively long tradition of applied research in this area has consistently established that complex and challenging jobs, as opposed to simplified, uninteresting ones, generally promote greater job satisfaction, higher work motivation, and, in some cases, lower incidences of anxiety, depression and other indices of psychological strain (Fraser, 1947; Kornhauser, 1965; Herzberg, 1966; Karasek, 1979; Wall and Clegg, 1981; Broadbent, 1985).

One important aspect of an individual's mental health concerns the level of psychological arousal, which can range from high to low, and the nature of the associated feelings, which can range from positive to negative. Thus, high arousal can be positive (for example, enthusiasm) or negative (for example anxiety), whilst low arousal can also be positive (for example contentment) or negative (for example depression) (see Warr, in press, for a fuller discussion of these issues).

In this context Specialist Control engenders lower levels of psychological arousal than Operator Control simply because the operators have fewer problems to solve and less interesting work to do. Similarly, there are lower levels of arousal in predictable than in uncertain environments. The worst conditions, therefore, for sustaining positive feelings and motivation are represented by predictable environments organized using the first approach, whilst the best exist in uncertain environments organized by the second. In the worst case, psychologically speaking, Specialist Control will be accompanied, amongst operators, by feelings ranging from contentment through apathy to depression. At the other extreme, Operator Control in an uncertain environment, feelings will range from enthusiasm to anxiety. These predictions are summarized in Figure 4.

Where other factors allow operators to exercise their skills to solve production problems with some degree of success (for example they have the skills, the production targets are reasonable and the production system is well managed and supported), then positive feelings will be more common than negative ones.

More generally, the Specialist Control approach results in simplified operator jobs, known to have negative psychological effects, whereas the Operator Control method involves enriched jobs of established benefit to operators' mental health.

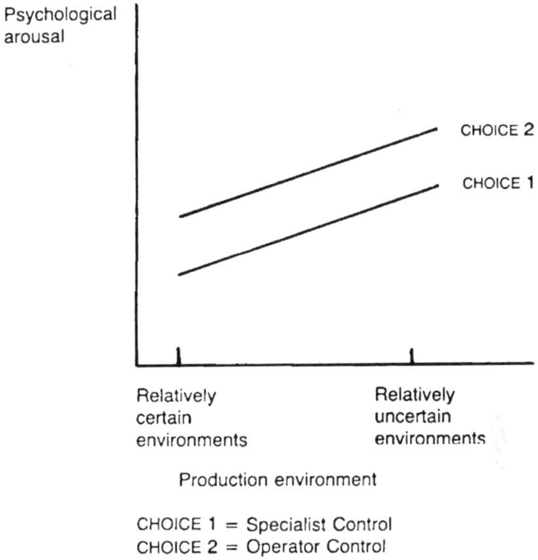

Figure 4. Choice and psychological arousal.

Remarks

This section has considered two different choices regarding the management and operation of CNC machine tool working. The Specialist Control choice is equivalent to deskilling and elsewhere has been called the 'operator monitor' approach, whilst the Operator Control choice is equivalent to enskilling and has been labelled the 'operator midwife' approach (see Wall, 1986). The choices are summarized in Table 1.

Three further comments are appropriate. First, the choice between these two approaches to control is not limited to CNC working, but applies more generally to other forms of FAS. Second, it is apparent that the typical practice of adopting a sequential approach to the design of FAS, where technical considerations take priority, means that these managerial choices and problems cannot be addressed properly. In practice these issues are very complex and should form the basis of extensive discussion and consultation with the interested parties throughout the planning and design of FAS. This effort is worthwhile since the choices are important both for the people concerned and for the overall performance of the system.

Table 1. Two choices for managing and operating FAS

	CHOICE 1 Specialist Control	CHOICE 2 Operator Control
Philosophy and objectives	Control by specialists Maximize management control	Control by operators Maximize responsiveness and system utilization
Method and structure	Specialize using a differentiated structure	Multi-skilled operators with simple structure
Personnel plans	Training for indirect staff Low grading and pay for operators IR issues between groups	Training for operators High grading and pay for operators
Benefits and costs	e.g. Low direct labour costs and 'tight' management control, but high indirect costs and poor utilization	e.g. Low indirect costs and good utilization, but high direct labour costs and lack of management control
Circumstances	e.g. Large factory, bureaucratic organization, big batches, simple products	e.g. Small factory, organic organization, small batches, complex products
Performance and mental health	Relatively low performance in uncertain environments Relatively poor mental health	Relatively high performance in uncertain environments Relatively good mental health

More generally, human factors, organizational and other management criteria should be introduced at the design stage of FAS, to ensure that the design process does not exclude or unduly constrain effective forms of operation and management (see Rosenbrock, 1983; Clegg and Corbett, in press).

And finally, it should be noted that Specialist Control represents the most common, as opposed to the most effective, approach to the management and operation of FAS (see Winch, 1983).

THE FUTURE

It was pointed out earlier that the companies with experience of FAS are in the first 'stand-alone' evolutionary period of advanced manufacturing technology. But many commentators agree that three developments can be expected over the next few years (see, for example, Northcott, Fogarty and Trevor, 1985). In the first place, technical innovations will enable FAS to become more integrated and more complex. It is already evident from those companies with experience of FMS that such production environments can be very complex and not necessarily more flexible (see Gerwin and Leung, 1980; Clegg and Corbett, in press).

Second, such technologies will increasingly be applied to small batch manufacturing environments where there is enormous potential for significant improvement through integrated control. For example, it is not uncommon in small batch engineering environments for it to take eight weeks to make a product that actually requires only a few hours machining time. For the rest of the time, the product is queuing. Given that around 70 per cent of engineering products fall into the medium batch, medium product range category (NEDO, 1985), there will undoubtedly be fierce competition through the application of FAS in this market.

A third trend is that companies will increasingly compete through flexibility. Thus they will compete by trying to respond faster to market demands and indeed by encouraging the market to demand more choice. Taken together these three trends indicate that FAS will get more complex and integrated, and will increasingly be applied to small batch production markets which themselves will become more common and more competitive. FAS will thus be applied to (and may result in) increasingly uncertain production environments according to the criteria defined earlier. The best managerial strategy will be to adopt and develop the approach labelled Operator Control, since this will benefit both system performance and operator mental health (Cummings and Blumberg, in press). This is particularly true in uncertain production environments, but elsewhere it also has the advantage of increasing the flexibility and responsiveness of a company through its investment in operator skills.

The argument is thus the reverse of the deskilling thesis, which argues that managers apply new technology to maximize their control of the production process. In practice either could happen, but we hope that the logic of effective and psychologically healthy production will prevail over the logic of management control.

CONCLUSIONS

Most manufacturing companies in the UK are in the first stages of experiencing computer-based Factory Automation Systems, and it is apparent that most research and development work in this area has displayed a myopic concern with the technical aspects of these new technologies, to the unfortunate exclusion of the managerial and human issues. These managerial concerns are complex, but effort in these areas has four potential pay-offs. It can lead to better system performance; it may result in more widespread acceptability; it may result in better mental health for the people working in the system; and it may improve a company's overall competitiveness.

A major issue concerning the application of computer-based FAS over the next few years concerns the ways in which they are and should be managed. More resources should be allocated to these issues, which will become more critical as the systems become integrated and are applied in more uncertain production environments. Critical choices face companies in how they operate and manage their FAS, and the adoption and development of an approach emphasizing Operator Control is advocated, since this is most likely to achieve the objective of promoting more effective and psychologically healthier production systems.

REFERENCES

Bessant, J. and Haywood, W. (1985) The introduction of Flexible Manufacturing Systems as an example of Computer Integrated Manufacturing. Unpublished paper, available from the Department of Business Management, Brighton Polytechnic.

Blackler, F. and Brown, C.A. (1986) Alternative models to guide the design and introduction of the new information technologies into work organizations. *Journal of Occupational Psychology, 59.*

Braverman, H. (1974) *Labor and Monopoly Capital: The Degradation of Work in the Twentieth Century.* New York: Monthly Review Press.

Broadbent, D.E. (1985) The clinical impact of job design. *British Journal of Clinical Psychology, 24,* 33–44.

Cherns, A.B. (1976) The principles of socio-technical design. *Human Relations, 29,* 783–792.

Child, J. (1984) *Organization* 2nd ed. London: Harper and Row.

Clegg, C.W. (1984). The derivation of job designs. *Journal of Occupational Behaviour, 5,* 131–146.

Clegg, C.W. and Corbett, J.M. (1986) Psychological and organizational aspects of computer-aided manufacturing: A review. *Current Psychological Research and Reviews, 5.*

Clegg, C.W. and Kemp, N.J. (1986) Information Technology: Personnel, where are you? *Personnel Review, 15(1),* 8–15.

Corbett, J.M. (1986) The design of machine tool technology and work: Technical science and technical choice. Sheffield: MRC/ESRC Social and Applied Psychology Unit Memo no 767.

Cummings, T.G. and Blumberg, M. (in press) Advanced manufacturing technology and work design. In T.D. Wall, C.W. Clegg and N.J. Kemp (eds) *The Human Side of Advanced Manufacturing Technology*. Chichester: Wiley.

Dodgson, M. (1985) *Advanced Manufacturing Technology in the Small Firm*. London: Technical Change Centre.

Fraser, R. (1947) The incidence of neurosis among factory workers. Industrial Health Research Board, Report no 90. London: HMSO.

Galbraith, J.R. (1977) *Organization Design*. Reading, Mass.: Addison-Wesley.

Gerwin, D. and Leung, T.K. (1980) The organizational impacts of flexible manufacturing systems: Some initial findings. *Human Systems Management, 1*, 237–246.

Herzberg, F. (1966) *Work and the Nature of Man*. New York: World Publishing Company.

Jones, B. (1984) Factories of the future, conflicts from the past. Paper presented at the seminar on Technology, Innovation and Social Change, University of Edinburgh. Mimeo available from the author, University of Bath.

Karasek, R.A. (1979) Job demands, job decision latitude and mental strain: Implications for job design. *Administrative Science Quarterly, 24*, 285–308.

Kemp, N.J. and Clegg, C.W. (1986). Information technology and job design: A case study on CNC machine tool working. Sheffield: MRC/ESRC Social and Applied Psychology Unit Memo no 716.

Kornhauser, A.W. (1965) *Mental Health of the Industrial Worker*. New York: Wiley.

Mueller, W.S., Clegg, C.W., Wall, T.D., Kemp, N.J. and Davies, R.T. (1986) Pluralist beliefs about new technology within a manufacturing organization. *New Technology, Work and Employment, 1*, 127–139.

NEDO (1985) *Advanced Manufacturing Technology: The Impact of New Technology on Engineering Batch Production*. London: National Economic Development Office.

Northcott, J., Fogarty, M. and Trevor, M. (1985) *Chips and Jobs: Acceptance of New Technology at Work*. London: Policy Studies Institute.

Northcott, J. and Rogers, P. (1984) *Microelectronics in British Industry: The Pattern of Change*. London: Policy Studies Institute.

Northcott, J., Rogers, P. and Zellinger, A. (1984) *Microelectronics in Industry: Survey Statistics*. London: Policy Studies Institute.

Rosenbrock, H.H. (1983) Social and engineering design of an FMS. In E.A. Worman (ed.) *Cape '83: Part 1*. Amsterdam: North Holland Publishing.

Sorge, A., Hartmann, G., Warner, M. and Nicholas, I. (1983) *Microelectronics and Manpower in Manufacturing*. Aldershot: Gower.

Taylor, J.C. (1975) The human side of work: The socio-technical approach to work system design. *Personnel Review, 4*, 17–22.

Trist, E.L. (1978) On socio-technical systems. In W.A. Pasmore and J.J. Sherwood (eds) *Socio-Technical Systems: A Source Book*. La Jolla: University **Associates.**

Wall, T.D. (1986) Advanced manufacturing technology: The case for the operator midwife. Sheffield: MRC/ESRC Social and Applied Psychology Unit Memo no 775.

Wall, T.D. and Clegg, C.W. (1981) A longitudinal field study of group work design. *Journal of Occupational Behaviour*, *2*, 31–49.
Warr, P.B. (in press) *Work, Unemployment and Mental Health*. Oxford: Oxford University Press.
Winch, G. (1983) *Information Technology in Manufacturing Processes: Case Studies in Technological Change*. London: Rossendale.

ERGONOMICS AND THE NEW TECHNOLOGIES

David J. Oborne

The problems arising from the application of new technology are not new. Ever since people first began to use tools to extend their own limited abilities – to help them to live and to expand their horizons – they have been confronted with the need to adapt both the facilities and themselves to create harmonious working environments. Archaeological evidence appears to suggest, for example, that stone-age man spent considerable time and effort adapting pieces of flint to make usable killing, digging and living implements. Even these hundreds of thousands of years ago, in designing their tools primitive people realized that to use the new technologies of the day with any degree of efficiency the tools needed to be designed and built to fit their own requirements; in this case their strength, dexterity and other physical capacities.

Modern new technologies have advanced considerably since then. These days we are able to generate new tools and applications, rather than relying on nature's provisions. These days the new technologies are mass-produced rather than being built for an individual purpose and for an individual user.

Unfortunately, the very technological advances that have brought humanity from the caves and provided us with safety, comfort and higher living standards have, in many respects, also brought with them the seeds of their own possible destruction. Since the introduction of mass-produced tools, over the centuries technological advances have not been matched with technological efficiency. Whenever new technology has been introduced – from bows and arrows towards the end of the last millenium to computers towards the end of the present one – the abilities required of the operators have generally been greater than the operators' abilities to perform. Only gradually, when people have learned to adapt themselves to the new technologies and when designers have learned to adapt their systems to the operators, have the new technologies been used with any efficiency. Unfortunately, by then the new technologies are no longer new. Other new technology is introduced and the cycle of inefficiency is continued.

This (hi)story has been, and still is being, repeated even with the new, information technologies that are presented to modern-day people. Systems are still being designed and built, controlling software is still being written, hardware is still being created, which, in the most part, requires from the operator more than the operator is able or willing to provide. Considerable evidence is available to support the argument that users' cognitive, perceptual and motor abilities are often not sufficient to operate successfully these new, highly sophisticated pieces of information processing machinery. The result is, then, that the theoretical capabilities of these systems are not, in actual fact, realized. Substantial amounts of time, effort and money are being wasted simply because man and machine are not in harmony.

Ergonomics and information technology

The importance of considering further the human operator's role within a technological system was made explicit with the development of the new discipline of ergonomics (sometimes called human factors) during World War II and after. Although the need to regard human abilities as variables that could affect operator efficiency was certainly understood for many years beforehand, it was only with the emergence of this new concept – that *all* features of the system (behavioural, physical and physiological) should be integrated and studied as a whole – that modern working situations began to be created in which people gained superiority over their environment. It was soon discovered, then, that by taking this comprehensive view of the working system – machine and operator working together and matching each other's abilities and requirements – efficiency could be increased to levels which were theoretically possible. (Oborne, 1982, 1985 provides examples of the extent to which costs have been reduced and efficiency increased through employing general ergonomics investigations.)

The central concept of ergonomics is the working system established between the machine and the operator. In terms of information technology systems, by linking 'man' and 'machine' in this way a relationship is created between these two components so that the computer presents information to the operator using its output mechanisms and via the operator's sensory apparatus to which a response is given. This time the operator uses his or her output apparatus – usually motor controlled – to present information via the computer's input systems. So, information is passed from the computer to the operator and back to the computer in a closed, information-control loop. Ergonomics' role is to investigate features of this loop and to make the information flow more efficient. Thus, aspects of the operator's performing behaviour – sensory, cognitive and motor abilities – are studied with the aim of discovering their strengths and limitations so that the system can be designed with these aspects in mind.

Ergonomics' role with the new technologies is to consider all aspects of the working relationship set up between the operator and the machine, in which, of course, the environment plays an important part. By gathering information concerning the strengths and weaknesses of the operator's perceptual, cognitive and motor abilities, and the extent that these features interact with the environment and the task, ergonomics is able to provide design parameters for the working system. By adapting the machine to fit the operator and by designing the system to suit these parameters, work can be carried out with maximum efficiency, comfort and safety.

Licklider (1960) emphasizes that, when applied to computers at least, this relationship can be described more as a symbiotic than as a 'master – slave' relationship (in whichever direction one conceives the relationship to work). Thus, he suggests that each of the two components in the system – 'operator' and 'computer' – depends on the other to perform well. The relationship, then, is not simply one in which the computer works to extend the abilities of the operator; nor is it one in which the operator simply reacts to stimuli provided by the computer. In many respects, the relationship is one in which not only does the computer depend on the operator to survive (this is obvious, at least at present) but also, in some ways, the operator is now unable to be a viable performing entity without the computer's abilities.

Shackel (1985) has argued cogently that taking account of the ergonomics aspects of IT is bound to become increasingly important as the application and use of IT changes. Thus, when IT was being developed and its uses probed over the past two decades or so, users – often dedicated users – were generally willing to put up with poorly designed interfaces since they knew and understood the system and its foibles. For a viable present and future IT industry, however, the new technology needs to be accepted by the users of the applications in addition to the designers. Doctors, lawyers, educationalists, managers, clerks can now chose between available systems. These 'end-users' are thus likely to become the individuals who will determine in the future whether or not a system or a piece of software will survive within a harsh economic climate.

> Thus, to be successful, the IT industry must improve the usability of interactive systems, and to do so the understandable orientation of designers in the earlier years must now be completely reversed. 'The last shall be first'; designing must *start* with the end user. Therefore the ergonomics aspects become paramount (p. 265).

Shackel also provides evidence for the economic impact of ergonomics. With personnel time and turnover these days costing more than the systems that they have to use, and the difference becoming greater, 'actions to reduce the human cost and simplify the human interface to computers will have the greatest impact on growth'.

This chapter provides examples of some of the different areas in which ergonomics investigations have enabled new, information technologies to be designed to suit the operator. It takes a selective view of the various aspects that ergonomics needs to consider to provide the relevant information for designers to design for usability. Because of the wealth of material to draw from and the limited space available some specific applications are not covered in detail – rather, the principles to be applied will be emphasized. Thus, neither speech input nor output is considered, nor the ergonomic problems surrounding new applications of IT in industry, such as flexible manufacturing systems or robotics.

Essentially, the approach is to consider the two sides of the information technology loop – presenting information from the machine to the user and vice versa. Within these two broad areas, evidence will be adduced to illustrate the need to consider the cognitive, perceptual and motor features of the user and the relationship of such data to designing various parts of the system.

PRESENTING VISUAL INFORMATION TO THE OPERATOR

Over the years, some considerable ergonomics work has been done to investigate many different features of display design. The thrust of such work can, essentially, be classified into two areas. First aspects of the display's physical parameters: brightness, contrast, colour, flicker, etc. and the extent to which these relate to the observer's abilities to see clearly the material presented. In the second area, work has been done to consider more the perceptual parameters of the material presented – character shape and size, display line spacing, character case, etc. Considerable amounts of data have arisen from such investigations and, since many of these results have been reported elsewhere (Cakir, Hart and Stewart, 1980; Oborne, 1985), detailed consideration will not be given here. Reasonably firm design parameters are thus quite widely available, so that there is no real excuse for designers and manufacturers to produce display systems which are less than usable for most people. Unfortunately, evidence is available which suggests that many of the guidelines are not being taken notice of.

Given that there appears to be a mis-match between the amount of design data available and the extent to which these data are incorporated into display designs, it is pertinent to question the reasons for the mis-match. A possible answer lies in the relevance of the available data to present-day applications. Without considering their application, ergonomics studies of, for example, optimum character founts remain just that – studies of optimum character founts. They are not studies of optimum character founts for spreadsheet use, for example, in which the display consists of high-density groups of numbers, nor of optimum character founts for use with graphs, in which the display is essentially pictorial and the characters present supplementary information. They are not even studies of optimum character founts for different sized display areas: 80 column, 40 column, high resolution, low resolution screens, for example. It is accepted, therefore, that some of the laboratory studies published in the available literature address questions that may not be central either to those being asked by designers or to those being faced by end-users. In this respect Shackel's criticisms of IT designers who frequently disregard the end-user might also be applied to IT investigators.

To produce relevant design data for presenting material to the user it is likely that future studies will have to relate more to the end-user and to the end-use. What does the material represent? How should it be presented to emphasize its meaning? What does the user wish to obtain from the material? Within ergonomics the need for such questions is generally becoming accepted – particularly with studies of task classification, task requirements and different forms of investigative methods.

Task classification

The simple example given above relating to the design of character founts indicates the importance of understanding something about the types of task normally carried out using a VDU screen and computer system. Different tasks may require different actions from the operator and impose different types and levels of strain. Yamamoto (1985), for example, points out that there is quite a considerable difference between normal, non-computerized, office work and office work that involves computer operations. Thus, whereas the non-computerized office worker generally works at a self-paced rate (for example, reading the next document when he or she decides to pick it up), the computerized worker's pace is generally determined by the computer – a response generally generates another display and the computer is ready for another response. Clearly, then, different working practices may require different ergonomics considerations. Indeed, evidence that even the same task – for example, reading a page – may be performed in a different way using a vertical (computer screen) as opposed to a horizontal (paper) display (Wright and Lickorish, 1983) suggests that far more consideration needs to be placed on the tasks performed by the operator, rather than just on the design of the system itself.

To produce a classification of the many different types of work carried out using visual displays would be a considerable task (Christie and Gardiner describe some of the approaches to such a task within this volume). However, some start has been made, although this has been primarily in the area of investigating health complaints of VDU users. Very little work has been done in considering other ergonomic aspects of the systems.

Grandjean (1980) classifies VDU work into two main types: data entry and conversational. In the first, the operator spends the largest proportion of the time entering data without much interaction with the machine: word processing, spreadsheet work, statistical analyses, etc. In conversational work, however, more interaction occurs with some time spent waiting for responses from the computer. Different physical and cognitive loads, then, are placed on the operator. Interestingly, the different types of task produce different patterns of fatigue complaints, with the data entry processors reporting higher frequencies of postural problems.

Coe *et al.* (1980) also reported variations in the effects of different working patterns on operator complaints (and thus efficiency). They divided the type of work done into four categories. Firstly, *input* work was similar to Grandjean's data entry work. Secondly, *creative* work which included tasks such as computer programming. Thirdly, *editing* tasks, which were similar to Grandjean's conversational category, in that continual interaction occured with the systems. Finally, Coe *et al.* included a *question–answer* category in which the tasks were highly screen interactive and involved a great deal of dialogue between the screen and keyboard. Again, significant differences were obtained between the classes of users for fatigue and strain complaints.

Task requirements and information layout

Following from classification of the types of task to be carried out is attention to the requirements of those tasks – what they expect from the operator and what the operator is able to give. Again, ergonomics considerations are paramount in this respect, since they stress the matching of operator and system by designing the system to suit the operator rather than vice versa.

Layout for information seekers. Wright (1986) argues strongly that task requirements must be viewed mainly in information processing terms. With such a view, she suggests, the material displayed needs to be compatible with the ways in which the operator is likely to process the information at a cognitive level. This means that presentation of the same type of material may have to be different in relation to the operator's needs, abilities and requirements.

Take, for example, a table of numbers within a spreadsheet and compare it with a table of numbers representing bus departure times. (Both of these, of course, may be presented either on paper or on a VDU screen.) Although both presentations provide numerical information in tabular form, they are likely to be used differently and may well require different forms of input. The spreadsheet user is likely to need the numerical information in order possibly to compare columns – profit and loss – to extract trends, etc. and will have the input data to hand in a particular format – finances, exam marks, etc. The timetable reader, on the other hand, will require that same sort of information – numbers – but for a different purpose. He or she will wish to know which bus to catch, at what time it arrives, whether there is a faster route, etc. A number of ergonomics studies are available which have considered the role of structure for different types of material to be presented.

It should be clear, then, that the ergonomics of the designed material presented to the user will need to vary in terms of what the task requires and in relation to the operator's processing and actions. In this case the advances provided by cognitive psychologists relating to perception, perceptual organization, thinking and reasoning should all play a role in determining the extent to which displayed material can be designed to be compatible with the observer's own conceptions and abilities.

Layout for information readers. As well as the need to consider the form in which the user requires the information, a fundamental question also relates to the presentation for ease of reading itself. Studies of eye movements during reading and the extent to which the reader's cognitive structure imposes different reading patterns on the material are particularly germane in relation to text layout and arrangement.

Since a stable image is only formed when both the eyes and the object are stationary, the movement of the eyes over the material during reading is characterized by a succession of fast movements (saccades) and stationary periods (fixations). The importance of understanding and analysing saccadic eye movements to the present discussion lies first in the fact that material presented that does not fit the saccadic movement is likely to be perceived less efficiently or, possibly, missed altogether. Secondly, saccadic patterns relate to the cognitive processing occurring within the perceiver (Rayner, 1977). Again, this indicates the need to consider the processes involved in the task and what the user wishes to obtain from the material.

The two important parameters of saccadic movements that relate to reading efficiency are the duration of each fixation and the number of fixations required. In this respect, the average length of a saccade appears to be about 2 degrees of visual angle (which conforms to about six 10-pitch or eight 12-pitch character spaces). For skilled readers the average fixation durations lie between 200 and 250 msec (Rayner, 1977), although there is a great deal of individual variability in this matter and this is thought to relate to reading skill.

Discussing the relationship of line saccades to the reading process, Bouma (1980) points to the importance of typographical design:

> The horizontal extent of line saccades is controlled by visual information in the left visual field, concerning the far left-hand margin, which should therefore be in a straight vertical line . . . with a sufficiently wide margin . . . The vertical extent of line saccades is controlled by perceived inter-line distance. If this vertical component is inaccurate, the eye may mistakenly jump over two or perhaps even three lines.

Although information regarding the eye's saccadic movements indicates the reader's control movements and possible times taken to process a piece of text, it is not the full story. Information is also needed concerning the amount of information that the reader can process at any one fixation. This is known as the 'span of perception' and is another aspect which determines reading speed and accuracy. Evidence from reading studies (McConkie and Rayner, 1975) suggests that the smaller the text area available to the subject, the longer it takes to read.

Furthermore, although material that appears central to the eye is most readily perceived and interpreted, readers also register information that falls outside the normal perceptual area. To what extent such material is used in comprehension rather than in eye control, however, is debatable. McConkie (1976), for example, has suggested that readers obtain different types of information from different regions within the perceptual span during a fixation in reading. Information falling on the fovea (the centre of the retina) is processed for its semantic content and information from the edge of the retina (the parafoveal area) is limited to rather gross featural information such as word shape and word length.

From an applied viewpoint, then, the importance of understanding the nature and parameters of saccadic eye movements during reading concerns the layout of textual and other material to fit in with these movements, and the division of words, phrases and sentences so that they are not frequently split in places that make comprehension difficult. Thus, the evidence that each saccade encompasses about six to eight character spaces should suggest that line lengths ought to be some multiple of this distance so that the eye does not need to make partial saccades. If the width is too small, however, only a few saccades may be possible on any one line, thus necessitating recursions to previous lines with the attendant problems of directing the eyes to the beginning of another line. On the other hand, if it is too large, too many saccades may be needed to scan the line.

More importantly, perhaps, it is necessary to ensure that when the eye is at the extreme right-hand end of one line it does not have too far to travel back to begin the next line (again, the optic control mechanism can lead the eye to the beginning of the wrong line). In this context, Bouma (1980) relates the length of a line of text (that is the distance between the left and right-hand margins) to the angle over which the eye travels to reach the next line. This, he suggests, should be approximately 2°. For long lines of text, therefore, the interline spacing should be reduced.

LINKING THE OUTPUT TO THE INPUT

Although considerable space has been given in this chapter to the importance of displaying material in a way that suits the cognitive capacities and wishes of the operator, the other side of the coin – the way in which the operator presents his or her information to the machine – should not be forgotten. This, after all, represents the other half of the closed-loop system so dear to the hearts of ergonomists and it is as important to understand the user's abilities and capacities in this respect as it is to understand them in terms of the material displayed. However, before considering input devices in any detail, the cognitive link between input and output actions should be discussed. That is, the extent to which the operator's actions need to be compatible with those of the display and vice versa. This is an area of ergonomics study called *compatibility*.

The need to ensure compatibility arises for three reasons. First, an incompatible display–control relationship is likely to lead to reduced response speeds. Second, the learning time for the operation of equipment on which the controls are compatible with the display will be much shorter than if they were incompatible. Third, and perhaps most insidiously, when placed under stress an operator's performance on equipment with incompatible display–control relationships will deteriorate as he or she reverts to the relationship expected to occur.

Three main ways of arranging compatibility between control and display exist. The first is *spatial compatibility*, which occurs when the position of items in the display suggests the appropriate control response. Second is movement compatibility. In this case the movement of items in the display suggests the way in which the associated control should be operated, and vice versa. For example, most operators would expect the right-hand cursor control key to move the cursor to the right of a screen, and the left-hand key to move it leftwards. Relationships which are expected by the majority of the population are described as population stereotypes. Oborne (1982) and Loveless (1962) provide details of these types of compatibility relationships and suggest means of predicting their direction. Last is cognitive compatibility, in which the actions required of the operator by the program itself need to be compatible with the user's expectations of their effects.

The importance of ensuring cognitive compatibility in computer programs and system commands has been highlighted by workers at the Applied Psychology Unit in Cambridge (Morton *et al.*, 1979;

Barnard *et al.*, 1981; Hammond *et al.*, 1981; Barnard *et al.*, 1982). In essence, they extract three forms of cognitive incompatibility: linguistic, memory and perceptual.

Linguistic incompatibility can occur at both a syntactic and at a semantic level. It often arises at a syntactic level because the information following commands such as DELETE, MOVE, INSERT, are generally used in abreviated forms, such as DELETE *x*, *y* or MOVE *x*, *y*. In 'natural' language, that is the language used by the operator, the above abbreviated commands might well be interpreted as DELETE (information *x*), from (file *y*), or MOVE (information *x*), to (file *y*). If this is how the computer programmer or system designer intended the actions to take place, then there is no syntactic incompatibility. Often, however, commands of the form DELETE *x*, *y* imply the reverse of what is expected in natural language, that is DELETE *y* from file *x*.

Semantic linguistic incompatibility can occur in a similar way. In this case, however, the problem arises over meanings of commands – particularly when the commands are computer-centric. Examples might include the use of the terms PUT and GET or LOAD and DUMP. Both of these pairs of operators are often used to transfer information from the computer to some storage medium and vice versa. However, the direction of the transfer is only immediately obvious if the user has already accepted that the computer is at the centre of the operation. Carroll (1982) also discusses some of these semantic incompatibility problems.

Memory incompatibility arises because the machine's requirements of the user's memory capabilities can be incompatible with the user's actual abilities. Again, linguistically incompatible terms can increase the memory load required because they require the user to remember, for each command, the relationship between the variables *x* and *y*.

Perceptual incompatibility relates primarily to the presentation of information as displayed on the computer screen and its relationship to the operations required of the user.

PRESENTING THE INFORMATION TO THE COMPUTER

Whereas the behavioural and ergonomic questions surrounding computer output – its design and presentation – relate essentially to the operator's cognitive capacities and requirements, inputting information from the operator to the computer concerns more the use of limbs. Thus, the considerations relate to the user's skill, and psychology's role in this respect is to understand and manipulate skilled behaviour. However, the importance of cognition should not be forgotten because it is a simple and obvious fact that the execution of a skill depends on the cognitive structure available to the operator. For example typing, a skilled behaviour that is very important to the efficient use of modern computer input devices, is seen to be as much a cognitive skill as it is a motor skill. In this case, considerable preprocessing of the material-to-be-typed occurs *before* the fingers hit the keys.

This part of the chapter considers various types of input device that are currently available, and the extent to which ergonomists have had a significant part to play in their design for effective use.

In general, the primary ways in which information is passed (input) to the computer is via one of the operator's three effector systems: limb movement and touch (usually using the hands or fingers), speech, and even eye movement. A number of possible devices can be conceived, although unfortunately there have been very few comparative studies to investigate which is the best type of control for particular circumstances. Those studies that have been reported have generally been restricted in their application and have considered simply the relative efficiencies of different controls for the simple task of selecting an item from a screen. Furthermore, no study has been reported that has considered operator preference. The work which has been done, however, has consistently indicated keyboard controls to take longer and to be more prone to errors (Earl and Goff, 1965; Goodwin, 1975). For inputting discrete pieces of information, continuous positioning devices such as lightpens are the most efficient (Card, English and Burr, 1978).

Keyboards

(This overview of research on keyboard design and input devices also appears in 'Psychology Survey 6', edited by H. Beloff and A.M. Colman, published by The British Psychological Society, 1987.)

Alphabetic keyboards. The normal QWERTY typewriter keyboard has been in existence since before the beginning of this century and was designed to conform to the mechanical constraints of contemporary typewriters: the apparent haphazard arrangement of letters was developed to slow typists down to prevent jamming of the keys. This presents a prime example of how systems can develop and become accepted .within new technology for entirely the wrong reasons. Ergonomics needs constantly to expose inappropriate design and, possibly more importantly, inappropriate use of design.

Despite the apparent inappropriateness of the QWERTY arrangement it would appear to have some saving graces – particularly from the viewpoint of the physical loads placed on operators' hands. It does distribute evenly the workload assigned to each hand and thus may reduce fatigue. Noyes (1983a), for example, argues that common letter sequences typed on the QWERTY board involve either alternate hands being used, the whole hand being moved over the keyboard or non-adjacent fingers being moved sequentially.

A number of alternative keyboard arrangements have subsequently been proposed. All are based on the frequencies with which letters and letter pairs occur in the English language. The two which have captured most experimental time are the Dvorak and the alphabetic board.

The Dvorak board (patented by A. Dvorak in 1932) was produced as a result of a decade of physiological and language research. The essential feature of the key arrangement is that all vowels and the most used consonants are on the second (or 'home') row, so that something like 70 per cent of common words are typed on this row alone. Generally, the arrangement means that vowels are typed with the left hand and frequent (home row) consonants with the right hand, producing, it is argued, a more even distribution of finger movements and a bias towards the right hand. It also reduces the between-rows movement by 90 per cent, and allows 35 per cent of all words normally used to be typed on the middle row.

Controversy presently exists as to the relative merits of the QWERTY and the Dvorak boards. For example, a US government sponsored study in 1956 demonstrated little difference between the arrangements (Alden, Daniels and Kanarick, 1976). Martin (1972), however, discusses (unreported) novice training experiments carried out in Great Britain which demonstrated a 10 per cent saving in training time using the Dvorak board. Furthermore, Dunn (1971) argues that the Dvorak board is superior in terms of ease of learning, reduced likelihood of error and fatigue and increased speed of entry.

On the alphabetic board, keys are arranged as the name suggests: from A to Z. The argument behind the use of this arrangement is, quite simply, that an alphabetical ordering of the keys makes logical sense, particularly to inexperienced typists, who need to spend considerable time learning the QWERTY arrangement.

Despite the apparent logic of using an alphabetically arranged board, Norman and Fisher (1982) point out that the available studies do not support the view that inexperienced typists find the alphabetic board easier to use. Indeed both Hirsch (1970) and Michaels (1971) have shown that for semi-skilled typists, keying rates and error correction are better using the QWERTY board, and performance on the two boards is essentially the same for novices. Norman and Fisher suggest two reasons for these findings: First an experimental one: it is difficult to find subjects who have not had some exposure to the QWERTY arrangement. Second, the alphabetic keyboard, although logically superior, still requires considerable visual search and mental processing (to remember, for example, that 'm' appears after 'k'). At the novice stage at least, therefore, all keyboard layouts are equivalent. Once the skill has been learned visual feedback gives way to more efficient feedback from the limbs themselves, so that the different board arrangements are likely to be equally efficient.

Numeric keyboards. Fewer studies have been performed to determine the optimum arrangement of the numeric keys (i.e.0–9) than the alphabetic keys – possibly because, with only 10 keys, there are fewer sensible arrangements that can be accommodated. A number of these arrangements were investigated by Deininger (1960) in a study of pushbutton telephone sets. Four designs were shown to be roughly equally acceptable on criteria such as keying time, errors and 'votes' for and against. For 'engineering' reasons, however, Deininger suggested an arrangement of a 3+3+3+1 matrix starting with 1,2,3 on the top row and ending with 0 below the third row. Indeed, the 'standard' telephone keypad has this arrangement.

Although this arrangement has become standard for telephone keypads, it is not currently used for numerical input on keyboards such as calculators. This is normally the reverse of the telephone arrangement, the keys on the 3+3+3+1 matrix having the order 7,8,9; 4,5,6; 1,2,3; and 0. Conrad and Hull (1968) compared the keying efficiency of these two types of arrangement. No significant differences were obtained in terms of the speed of data entry but they did find that significantly fewer errors were made using the telephone keypad (1,2,3; 4,5,6; etc.) than with the calculator pad arrangement (7,8,9; 4,5,6; etc.) 6.4 per cent versus 8.2 per cent).

Chord keyboards. In the search for improved ways of keying data, particularly alphabetic data, the possibility of reducing the number of keys by requiring the operator to press more than one key at a time has often been suggested. Such key arrangements are called chord keyboards and they appear in many different forms. (Litterick, 1981, describes some of these boards, and Noyes, 1983b, describes the history and development of chord keying.) The efficiency of such boards, of course, will be determined by the combinations of keys used to produce particular letters – from the viewpoint both of the operator's ability to use various finger combinations and to learn and remember key sets (see Seibel, 1964).

Very few experiments have been performed to compare directly keying performance using a typewriter and a chord keyboard. Again, this is probably because of difficulties in obtaining matched groups of subjects and being able to train them for very long periods of time using the same instructor. Nevertheless, the comparative studies that are available have demonstrated a chord keyboard performance superiority (Bowen and Guiness, 1965) and reduced training time (Conrad and Longman, 1965).

Other types of input device

With the need to input varying forms of information, computer systems have caused a number of innovative designs to be created. Unfortunately, as will become apparent, the amount of behavioural and ergonomics input to the design of such devices has been minimal. Consequently, such devices are appearing more frequently on the market without having been considered in terms of user ability. As argued above, history is repeating itself. This time, however, users are unlikely to have to wait for nearly a century before the usable features of such devices are studied – as they did for evaluations of the QWERTY keyboard. This time, market forces will not allow it, since only the ergonomically designed devices – designed to fit the user and users' actions – will survive. The remainder of this section describes some of these devices and indicate areas of study that are sadly lacking.

Touch displays. These allow the user to input information to the machine simply by touching an appropriate part of the screen or some representation of the screen. Since the computer screen both presents information to and receives information from the operator, it combines the functions of keyboard and display. Both Hopkin (1971) and McEwing (1977) discuss the advantages of screen based displays, which can be summarized thus: they are easy and fast to use, training time is reduced (Usher, 1982), they minimize errors, they are flexible, and operator reaction is generally favourable. Against these advantages, however, Pfauth and Priest (1981) suggest a number of disadvantages: initial high cost for the system, increased programmer time, reduced flexibility for some types of input, possible screen glare, physical fatigue from reaching to the screen, and the finger and hand blocking the operator's line of sight to important areas of the screen.

Light pens. Like touch displays, lightpens are fully interactive control devices. They can be used effectively to position the cursor on the screen or to select responses from a 'menu' displayed to the operator. Unfortunately, little research appears to have been carried out to investigate either the design or the efficiency of this type of control, although Oborne (1985) discusses various features that should be important.

Bar code scanners. These are devices which both look and operate very much like lightpens, but they are not used interactively with the computer screen; rather they are passed over alternate black and white bars, the composition of which contains the information to be input. They have a major advantage in that their operating postures are not constrained by the computer system itself so that the arm and hand do not need to be maintained under static load to enable the pen to touch the screen. However, Wilson and Grey (1983) point out that the fixed 'pen' system of scanners, in which the material to be read is passed over the scanner, can create postural difficulties for the operator.

Levers and joysticks. The difference between a lever and a joystick is simply that joysticks operate in two dimensions whereas levers only operate in one. For this reason, joysticks are used more often for cursor positioning. Because they are used in situations in which precision adjustments are made, it is desirable that only the hand and fingers are used, since these muscles are more densely supplied with nerves than, for example, the arm. For this reason joysticks are generally smaller than levers. To aid precision they should have resistance in all directions with, perhaps, a return to centre position if the hand is moved. Morgan *et al.* (1963) further suggest that the joystick should be designed to enable the operator to rest the wrist while making the movements, and that the pivot point should be positioned under the point at which the wrist is rested.

The roller ball and mouse. As the name suggests, 'roller balls' are spherically shaped objects which the operator can rotate in any direction. Their distinctive characteristic is that they rotate within a socket; thus they are fixed pieces of equipment. The 'mouse', on the other hand, operates in a similar fashion to the roller ball but it is not fixed; the operator is able to move it around, much like a pen is moved around paper to form characters. Card, English and Burr (1978) have demonstrated the superiority of these input devices over the conventional keyboard when used to move a cursor around the screen.

SUMMARY

Through various examples, this chapter has emphasized the need to consider all aspects of the operator's cognitive, physical and physiological behaviour when interacting with computer systems. It has stressed the importance of understanding the many different uses to which computers can be put and adapting the hardware and software accordingly. Only when this occurs, when the user and the system are in harmony with each other, when a true symbiotic relationship can emerge, will users be able to apply the computer to their and its fullest potential. Since the users are the choosers, only when this occurs will current and potential computer systems become competitive and viable.

REFERENCES

Alden, D.G., Daniels, R.W. and Kanarick, A.F. (1976) Keyboard design and operation: A review of the major issues. *Human Factors, 14*, 275–293.

Barnard, P.J., Hammond, N.V., Morton, J., Long, J.B. and Clarke, I.A. (1981) Consistency and compatibility in human–computer dialogue. *International Journal of Man–Machine Studies, 15*, 87–134.

Barnard, P.J., Hammond, N.V., MacLean, A., and Morton, J. (1982) Learning and remembering interactive commands. IBM research report HF 055. Portsmouth: IBM.

Bouma, H. (1980) Visual reading processes and the quality of text displays. In E. Grandjean and E. Vigliani (eds) *Ergonomic Aspects of Visual Display Terminals*. London: Taylor and Francis.

Bowen, H.M. and Guiness, G.V. (1965) Preliminary experiments on keyboard design for semi-automatic mailsorting. *Journal of Applied Psychology, 49*, 194–198.

Cakir, A., Hart, D.J. and Stewart, T.F.M. (1980) *Visual Display Terminals*. Chichester: John Wiley.

Card, S.K., English, W.K. and Burr, B.J. (1978). Evaluation of mouse, rate-

controlled isometric joystick, step keys, and text keys for selection on a CRT. *Ergonomics, 21*, 601–613.
Carroll, J.M. (1982) Learning, using and designing command paradigms. *Human Learning, 1*, 31–62.
Coe, J.B., Cuttle, K., McClellon, W.C., Warden, N.J. and Turner, P.J. (1980) *Visual Display Units*. Report W/1/80. Wellington: New Zealand Department of Health.
Conrad, R. and Hull, A.J. (1968) The preferred layout for data-entry keysets. *Ergonomics, 11*, 165–173.
Conrad, R. and Longman, D.J.A. (1965) Standard typewriter versus chord keyboard – an experimental comparison. *Ergonomics, 8*, 77–88.
Deininger, R.L. (1960) Human factors engineering studies of the design and use of pushbutton telephone sets. *The Bell System Technical Journal, 39*, 995–1012.
Dunn, A.G. (1971) Engineering the keyboard from the human factors viewpoint. *Computers and Automation*, February, 32–33.
Earl, W.K. and Goff, J.D. (1965) Comparison of two data entry methods. *Perceptual and Motor Skills, 20*, 369–384.
Goodwin, N.C. (1975) Cursor positioning on an electronic display using lightpen, lightgun, or keyboard for three basic tasks. *Human Factors, 17*, 289–295.
Gould, J. (1982). Writing and speaking letters and messages. *International Journal of Man-Machine Studies, 16*, 147–171.
Gotlieb, C.C. and Borodin, A. (1973). *Social Issues in Computing*. New York: Academic Press.
Grandjean, E. (1980) Ergonomics of VDUs: Review of present knowledge. In E. Grandjean and E. Vigliani (eds) *Ergonomics Aspects of Visual Display Terminals*. London: Taylor and Francis.
Hammond, N.V., Long, J.B., Morton, J., Barnard, P.J. and Clark, A. (1981) *Documenting human–computer mismatch at the individual and organisational levels*. IBM research report HF 040. Portsmouth: IBM.
Hirsch, R.S. (1970) Effects of standard versus alphabetical keyboard formats on typing performance. *Journal of Applied Psychology, 54*, 484–490.
Hopkin, V.D. (1971) The evaluation of touch displays for air traffic control tasks. IEE Conference on Displays: Publication no. 80
Licklider, J.C. (1960) Man–computer symbiosis. *Institute of Radio Engineers Transactions of Human Factors in Electronics*.
Litterick, I. (1981) QWERTYUIOP – dinosaur in the computer age. *New Scientist, 89*, 66–68.
Loveless, N.E. (1962) Direction-of-motion stereotypes: A review. *Ergonomics, 5*, 357–383.
Martin, A. (1972) A new keyboard layout. *Applied Ergonomics, 3*, 48–51.
McConkie, G.W. (1976) The use of eye-movement data in determining the perceptual span in reading. In R.A. Monty and J.W. Senders (eds) *Eye Movements and Psychological Processes*. Hillsdale, N.J.: Lawrence Erlbaum Associates.
McConkie, G.W. and Rayner, K. (1975) The span of effective stimulus during a fixation in reading. *Perception and Psychophysics, 17*, 578–586.
McEwing, R.W. (1977) Touch displays in industrial computer systems. In *Displays for Man–Machine Systems*. London: IEE.

Michaels, S.E. (1971) Qwerty versus alphabetic keyboards as a function of typing skill. *Human Factors, 13*, 419–426.

Morgan, C.T., Cook, J.S., Chapanis, A and Lund, M. (1963) *Human Engineering Guide to Equipment Design*. New York: McGraw-Hill.

Morton, J., Barnard, P., Hammond, N. and Long, J.B. (1979) Interacting with the computer: A framework. In E. Boutmy and A. Danthine (eds) *Teleinformatics '79*. Amsterdam: North Holland.

Norman, D.A. and Fisher, D. (1982) Why alphabetic keyboards are not easy to use: Keyboard layout doesn't much matter. *Human Factors, 24*, 509–519.

Noyes, J. (1983a) The QWERTY keyboard: A review. *International Journal of Man–Machine Studies, 18*, 265–281.

Noyes, J. (1983b) Chord keyboards. *Applied Ergonomics, 14*, 55–59.

Oborne, D.J. (1982) *Ergonomics at Work*. Chichester: John Wiley.

Oborne, D.J. (1985) *Computers at Work: A Behavioural Approach*. Chichester: John Wiley.

Pfauth, M. and Priest, J. (1981) Person–computer interface using touch screen devices. *Proceedings of the 25th Annual Meeting of the Human Factors Society*. Baltimore: HFS.

Rayner, K. (1977) Visual attention in reading: Eye movements reflect cognitive processes. *Memory and Cognition, 5*, 443–448.

Rayner, K. (1978) Eye movements in reading and information processing. *Psychological Bulletin, 85*, 618–660.

Seibel, R. (1964) Data entry through chord, parallel entry devices. *Human Factors, 6*, 189–192.

Shackel, B. (1985) Ergonomics in information technology in Europe – a review. *Behaviour and Information Technology, 4*, 263–289.

Usher, D.M. (1982) A touch sensitive VDU compared with a computer-aided keypad for controlling power generating plant. Paper presented to IEE Conference on Man–Machine Systems.

Wilson, J. and Grey, S. (1983) The ergonomics of laser scanner checkout systems. In K. Coombes (ed.) *Proceedings of the 1983 Ergonomics Society*. London: Taylor and Francis.

Wright, P. (1986) Phenomena, function and design. In D.J. Oborne (ed.) *Contemporary Ergonomics 1986*. London: Taylor and Francis.

Wright, P. and Lickorish, A. (1983) Proof-reading texts on screen and paper. *Behaviour and Information Technology, 2*, 227–235.

Yamamoto, S. (1965) A study of VDU operators' information processing based on saccadic eye movement and response times. *Ergonomics, 28*, 855–868.

OFFICE SYSTEMS

Bruce Christie
and
Margaret M. Gardiner

Office systems is likely to be the single largest sector in information technology during the late 1980s and through the 1990s. It is an area in which over the last several years increasing recognition has been given by industry to the potential contributions that psychologists can make. In this chapter, we focus on three areas in particular:

- The analysis of functional requirements – where psychologists can help in the definition of what services a new product could usefully offer its intended users.
- The development of appropriate models of user behaviour – where psychologists can provide insights into human information processing and other aspects of human behaviour in relation to user–system interaction.
- Design and evaluation of new products – where psychologists can bring to bear a wide repertoire of methods developed in other areas of psychological research.

We do not discuss how the psychologist can help in the analysis of any one particular organization's needs or in the introduction of new systems into a particular organization. These are areas where psychologists can indeed make important contributions but they are covered elsewhere in this volume, especially Chapter 2 ('Management, organizations and the new technologies') and Chapter 10 ('Attitudes to information technology'). Also, we focus on the contributions that psychologists can make in the newer field of user-interface design (emphasizing the design of the software) rather than the older field of traditional ergonomics (emphasizing the hardware and workplace) which is covered in Chapter 4 ('Ergonomics and the new technologies').

ANALYSIS OF FUNCTIONAL REQUIREMENTS

A common reason why many office products have failed is that they have been designed without sufficient attention to the needs of their intended users. The products may have been interesting in terms of their specific features – but they did not provide the right sort of 'services' to help office workers.

Psychologists can make a valuable contribution to the development of new, more appropriate products through the analysis of office work – both as it is today, and in terms of the underlying functions that may be served in different ways in an electronic environment. This kind of analysis can provide a basis for developing a high-level specification of what kinds of services a new product needs to provide. Put simply, it can help in the definition of *what* a product should do for its user – the *how* being the subject of a later stage in the design process (considered later in this chapter).

A number of different approaches is possible. Psychologists do not have the monopoly on them and some of the examples given here come from the work of people who might not consider themselves psychologists. The training which psychologists receive does mean that they are in a strong position to contribute to the range of methods available, select the particular method or combination of methods that is appropriate in a given case, and apply that method or methods professionally. The various particular approaches possible fall into two main broad categories: the empirical approach and the model-driven approach.

The empirical approach

The empirical approach emphasizes the need to collect data about the particular type of office environment for which the proposed product is intended. The aim is to understand through observation and analysis what work is done in the type of office concerned, and to infer opportunities for new products.

Time allocation studies. A common approach in this general category is to examine the way in which time is allocated in the office. For example, in an often-cited study by Engel *et al.* (1979) office principals were found to spend about 27 per cent of their time on activities clearly related to talking with people (for example telephone calls and meetings), and about 44 per cent of their time was spent on activities which were predominantly to do with papers (that is writing and reading); the remaining 29 per cent of their time was taken up by a variety of other activities (for example calculating, planning or scheduling, travelling, using equipment).

Such surveys can be useful as a first step in painting a 'broad-brush' picture of an organization's office activities, but they are very limited as a basis for inferring functional needs. One reason for this is that there is fairly wide variation in the results of such surveys, depending upon the particular organization concerned and the survey methods used. (See Doswell, 1983, for some examples.) The variation in findings makes it difficult to draw very general conclusions, although it has usually turned out that office workers spend more time working with papers than talking with people and that this difference is more pronounced the lower the grade of worker (for example many secretaries spend 80 per cent or more of their time working with papers). The amount of time often consumed by paper-work helps to explain the commercial success of word processing systems in recent years.

There is a further reason why such surveys are limited as a basis for inferring functional needs. There is no necessary correlation between the fundamental importance of any given type of activity and the amount of time currently spent on it. As Field (1985) has pointed out, in considering the findings from such surveys it is important not just to consider whether new technology could enable some activities to be done more efficiently, but to ask whether some activities need to be done at all; whether there are other activities that in principle are more important but which are hardly in evidence at the present time; and whether new technology can help to bring activities into new, more effective combinations.

'Comprehensive' studies. The way in which time is allocated to existing activities is only one kind of data that can be collected about office work. A wide range of other data can be collected relating more or less directly to office activities, information flows and organizational communication (Van de Ven and Ferry, 1980). Such data can include information about the design of individual jobs, the design of organizational units, aspects of inter-unit relations, and aspects of macro-organizational design.

Such 'comprehensive' analyses of an organization can never be truly comprehensive because the number of questions one could ask is, in principle, infinite. However, they can be useful for providing a more detailed picture of the type of organization for which an electronic office product may be intended than the narrower type of study based simply on allocation of time to different activities.

In the absence of a clear conceptual model of the office, it is difficult to be sure that even the most wide-ranging of such 'comprehensive' studies is actually collecting the most relevant or most useful data, or that some critical data are not being overlooked. The empirical approach therefore ultimately depends for its success on the development of useful models of the office, in order that data collection and analysis can be directed efficiently in an appropriate direction.

The model-driven approach.

In contrast to the empirical approach, the model-driven approach emphasizes the development of a clear conceptual model of the office, and then uses this as a basis for designing an office system. The utility of the model is then reflected in the success or failure of the office system in practice. The following two examples illustrate this approach.

Office work as procedures. Lum, Choy and Shu (1982) propose that office work can be considered to be a set of definable 'procedures'. These procedures involve the handling of 'forms'. What is meant by a form in this context is what one normally thinks of as a form, except that it also includes text documents (considered to be forms with very long data fields). They distinguish between the 'display form' (which is what the user sees) and the 'abstract form' (which is what the machine sees). The two are closely related, but the abstract form includes well-defined relationships between the fields of the form, which are often difficult to represent in the display form. Forms are considered to be hierarchical in structure; a particular field in a form may itself be described at the next level of detail as a form with several fields.

The definition of forms is done in some considerable detail, forming the basis for a language – called FORMAL (Forms ORiented MAnipulation Language). Office work is then regarded as a set of procedures made up of well-defined 'processes'. Each process takes as input one or more forms and normally produces one form as output. Lum, Choy and Shu explain how this model of the office has been used to build an experimental office system called OPAS at IBM Research Laboratories at San Jose. Other, commercially-available systems which emphasize the 'procedural' nature of office work include 'Ten-Plus' and 'Progress'.

Office work as problem-solving activity. Barber (1983) has used and extended ideas from artificial intelligence in order to describe office work as goal-directed problem-solving activity. He describes a language called OMEGA which is based on this view, and shows how it can be used to build office systems that support office work as problem-solving.

The 'goal' is a key concept in this model. In an OMEGA system, the office worker establishes a goal. This could be, for example, to send a message, or to complete a particular step in an office procedure. OMEGA then tries to achieve the goal. If it cannot do so, it notifies the user that the goal cannot be achieved or that contradictory information has been discovered during the attempt to achieve the goal. The user can then either modify the goal or provide further information that may allow the goal to be achieved.

OMEGA recognizes that goals in the real organizational world are often ill-defined, that what information is relevant to achieving a goal is not always clear, and that organizations involve many people whose activities may interact with one another in various ways that are difficult to predict. It is not intended to be a system which does the problem-solving for the human; it is a language which allows tools to be built which support the human in problem-solving rather than replacing the human.

Relationship between the two approaches

The two general approaches outlined above are not completely separate, but reflect a difference of emphasis. The empirical approach requires a conceptual model of the office in order to collect and analyse the appropriate data, and the model-driven approach needs empirical data in order to test the usefulness of the models. To date, however, it would seem that the model-driven approach has been the more fruitful in terms of direct contributions to prototype office systems.

MODELS OF USER BEHAVIOUR

Having decided on the general nature of the product concerned, in terms of what it is intended to help the user achieve, there comes the question of how to design the product in a way that is most suitable for the kind of person for whom it is intended. It is now widely recognized that a product aimed at the general office worker should not assume that the user is familiar with the kinds of concepts or has the kinds of skills that one might expect of a trained computer programmer or computer operator. The designer needs to be able to go beyond such basic considerations, however, and this is where psychological models of user–system interaction may in the future have something useful to offer.

Contributions from cognitive psychology

Most of the work to date has taken a cognitive view of the user – emphasizing the information-processing aspects of user–system interaction, and drawing on concepts from cognitive psychology and cognitive science.

Possibly the best-known example of this type of model is the GOMS model proposed by Card, Moran and Newell (1983), based at Xerox Parc. The four main concepts in the GOMS model are:

- ☐ *Goals,* which define the state of affairs to be achieved – they are organized hierarchically.
- ☐ *Operators,* which are elementary acts, necessary to create a change in the user's mental state and/or the task/system environment – they can be perceptual, motor or cognitive in nature.
- ☐ *Methods* for achieving the goals – these are conditional sequences of goals and operators that the user has already learned (not plans that are formulated at the time of executing the task).
- ☐ *Selection* rules for choosing among competing methods – these are of the form 'If such-and-such is true in the current task situation, then use method *m*'.

The GOMS model is a generic model which can be used to generate more specific models at different levels of analysis. It has been applied with some success to text-editing and other well-structured tasks, and to a semi-creative task (computer-aided circuit design) to predict performance (especially times taken). Card, Moran and Newell present some illustrative comparisons between different text-editing systems.

Contributions from social psychology

Although the emphasis in user–system modelling has strongly emphasized the cognitive aspects of user–system interaction, some researchers (for example Murray and Bevan, 1984) have argued that the social psychological aspects should not be overlooked. Consistent with this view, Richards and Underwood (1984) have shown that the social style of communication adopted by the electronic system can influence the social style adopted by the user. For example, subjects in their experiments were more likely to adopt a polite style themselves when the system addressed them with a polite, inexplicit introductory message than when other messages were used. They discuss how effects of this sort could be used to advantage in facilitating speech communication between the user and the system.

The social psychological view and the cognitive view of behaviour in this context need not be considered as competing, but can be seen to complement one another and offer concepts which may jointly enrich the view of user–system interaction. For example, the concept of behaviour as being organized according to a hierarchy of goals – evident in the cognitive GOMS model – is not inconsistent with a social psychological view. Such a view of behaviour is evident in the work of social psychologists such as Argyle and colleagues (for example Argyle, Furnham and Graham, 1981) and von Cranach and colleagues (for example von Cranach *et al.*, 1982). The von Cranach work is, in for fact, best considered as 'hybrid' in terms of its theoretical constructs, including both cognitive and social psychological concepts. The model which these researchers propose includes, for example, the 'cognitive' concepts of 'attention processes', 'goal determination' and 'cognitive control', but it also includes 'social psychological' concepts such as 'social control', 'values' and 'attitudes'.

Contributions from psychophysiology

Whilst both cognitive psychology and social psychology can make contributions to our understanding of user–system interaction, their areas of special concern have received less attention outside strictly research circles than that of a third area of psychology: psychophysiology. Whilst the non-psychologist with an interest in user–system interaction may not be very aware of the 'cognitive' or the 'social psychological' issues surrounding user–system interaction, (s)he will probably be aware of the debate concerning possible health hazards of VDUs. This debate is an example of the kind of area in which the psychophysiologist works.

Psychophysiology recognizes that humans working with electronic systems do not just 'perform tasks', in the sense of completing them within a particular time and with a particular level of accuracy or quality. They also respond physiologically (one concern of physical ergonomics – see Chapter 4 – but it goes beyond considerations of the physical ergonomics of the situation), and they have subjective experiences (some, but not all of which relate directly to the cognitive processes that are the concern of cognitive psychological models). Psychophysiology is therefore concerned with the whole situation, not just one aspect of it, and in particular, as Gale (1973) has suggested, psychophysiology is concerned with the integration of three domains: overt behaviour, subjective experience and physiology.

The nature of the linkages between the three domains can be debated but for practical purposes the Gale and Edwards (1985) view that changes in any one can affect either or both of the other two is probably the safest view to take.

The three domains can be regarded as sub-systems within the user of the electronic office system, the user in turn being a component of broader situational, organizational and societal systems. The sub-systems are all in continuous interaction, their relative dominance within the hierarchy varying according to circumstances. Overlaid on this, people vary in terms of which of their sub-systems tends to be the most dominant in most situations. As recognized by general systems theory, such systems have emergent properties – the whole (of the system) being greater than the sum of its parts (sub-systems).

The fact that systems have important emergent properties means it can be difficult to make predictions about how successful (in terms of user–system interaction) a new electronic office system will be, based only on consideration of one particular aspect of the system in isolation (for example time taken to complete a task as a function of the screen layout). It is necessary to deal with the complexity on its own terms, and not throw it away by dealing with only isolated components.

The degree of complexity involved in user–system interaction means that psychological research aimed at developing useful models of user–system interaction needs to be conducted in contexts that are 'ecologically valid' (representative of real-life situations) in preference to the highly simplified situations of the traditional experimental laboratory.

USER-INTERFACE DESIGN AND EVALUATION

Once it has been decided *what* a new office product should offer its user, there is a need to consider *how* it should do so. There are technical aspects to this question which are not of concern to the psychologist, and there are aspects of the design of the user-interface which are.

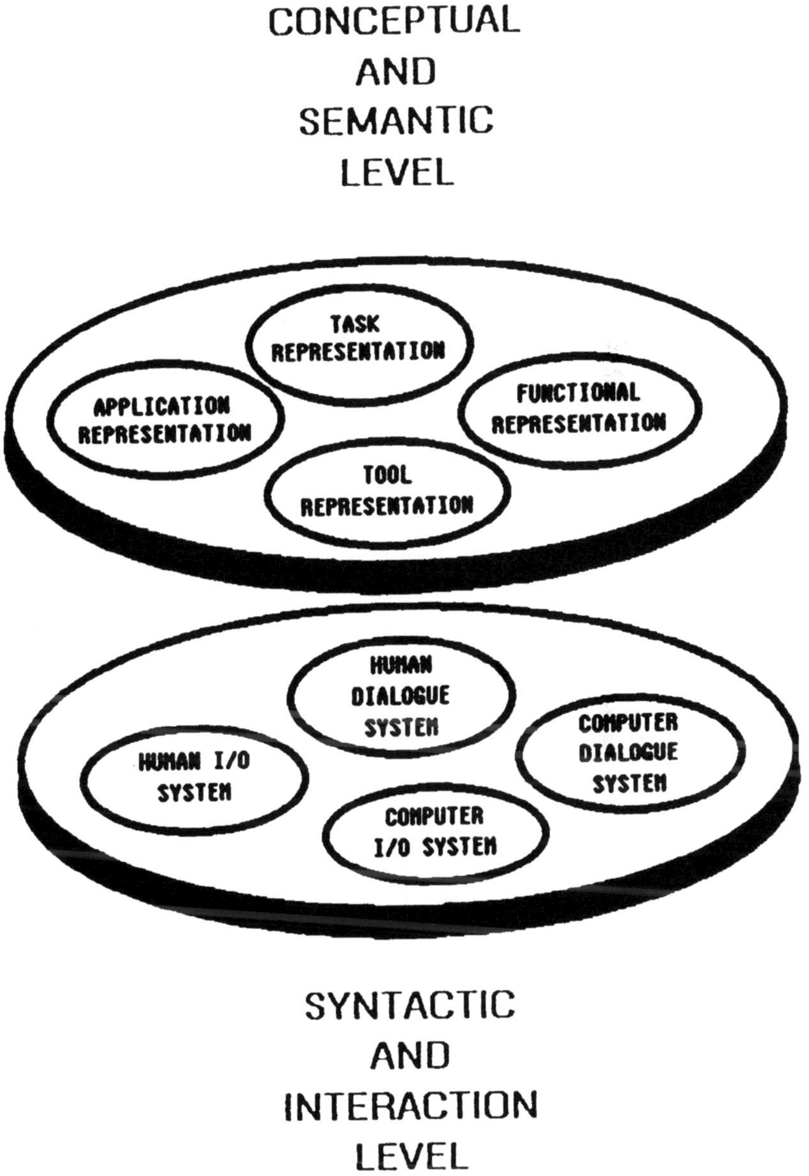

Figure 1. Levels of the user–system interface.

The user-interface is the product as it is presented to the user. Recently, work carried out under the ISO 'umbrella' has identified four separate 'levels' to the user–system interface (see Fähnrich and Ziegler, 1984, and Figure 1 in this chapter):

- a conceptual level, which deals with the mapping between the user's and the system's representation of the tasks the system can be used to achieve
- a semantic level, which addresses, amongst other things, the system tools and procedures which allow achievement of the overall task
- a syntactic level, to do with the actual structuring of the dialogue
- an interaction level, which addresses the actual input and output devices (human and electronic) and the physical aspects of hardware and software design.

There are also ancillary aspects of the design of electronic office systems which are not to do with the user-interface itself (in the sense defined here) but which are nonetheless important. These include user guidance materials, training, and so forth. Some of these are considered later in this chapter in connection with the implementation of office systems.

Design and evaluation are considered together because they normally go hand in hand. Even in something as apparently straightforward as the use of design guidelines, the design team behaves proactively (in reading/remembering guidelines and attempting to design in accordance with them) and retroactively (in judging to what extent a particular design – as it develops – is consistent with relevant guidelines).

Psychologists make contributions to the design and evaluation process in many ways, of which the following are illustrative rather than comprehensive.

As an expert on the design team

Design of office systems is normally done by teams, not by individuals. The teams concerned are multidisciplinary, and psychologists are playing an increasingly important part in such teams. Three important shifts are discernible in connection with this over the last several years:

- □ The expert concerned is increasingly likely to be invited to join the design team as a full member of the team, on a par with specialists from other disciplines (computer science, electronics), rather than to be used as a consultant who is only brought in occasionally when the design teams sees fit, and who has less influence than the members of the team proper.
- □ The person concerned is increasingly likely to have a background in cognitive psychology. Specialists in traditional ergonomics are still likely to be used on a consultancy basis, and to contribute primarily to the physical ergonomics of the workstation or other aspects of hardware layout.
- □ The advice of the psychologist concerned is increasingly likely to be sought early on in the design process – often at the very earliest stages when the product is only defined in the most general terms – rather than being sought only after all the critical design decisions have been made.

Having said this, there is still a long way to go before psychologists are contributing to the fullest extent to which they are capable. And it must be said that to some extent this depends upon psychologists being much more willing than they often are to 'come off the fence' and be prepared to make a mistake by saying something definite based on their best judgement at the time – rather than avoiding mistakes (and positive contributions to the design) by refusing to be specific. Psychologists too often behave as if the design process will stop (or should stop) 'until adequate research is done', when in fact the world moves on even if the psychologists concerned decline to be a part of it.

Through the provision of design guidelines

The success of design guidelines has been mixed, but it is certainly the case that in some organizations the design teams welcome concise statements of important 'dos' and 'don'ts' relating to the design of office products. These can be at their most successful when they are provided as concise 'manuals' rather than as texts on psychology or ergonomics. That is, the design team often appreciates having a document which it can use as a reference book, rather than a text to aid in learning about human factors or psychology. This is not to say that training is not important – it is, and is considered later in this chapter – but guidelines may be most likely to succeed when they are presented as a set of recommendations to follow, complementary to rather than as a vehicle for training. The design team is then more likely to use them in a similar way to standards, whilst recognizing their more tentative status and using them accordingly (that is as *guides* rather than rigid rules).

Design guidelines in this area have traditionally been based on a mixture of research and practical design experience – mostly the latter – and have focused largely on traditional ergonomic aspects. Specific guidelines can be found in a wide range of journal articles and other publications, but there have also been a number of attempts to bring these together in a more convenient form. These collections have often been distributed on an in-house, confidential basis, but a number are available publicly (Cole, Lansdale and Christie, 1985; Kidd, 1982; Smith, 1982; Galitz, 1980).

More recently, especially under the European Strategic Programme of Research and Development in Information Technology (ESPRIT) funded by the Commission of the European Communities and European Industry, there have been attempts to develop guidelines specifically based on the findings from research on human cognitive psychology (for example Marshall, Nelson and Gardiner, in press).

Through the provision of generic designs

Generic designs provide concrete illustrations of how to apply design guidelines. They show by example. Such a design includes design concepts that have a useful degree of generality, so they can be incorporated into many particular designs used across a range of products. The concepts incorporated into a generic design might include, for example: particularly good ways of allowing the user to scroll through a document; particularly interesting windowing techniques; particularly helpful ways of allowing the user to edit voice messages; and so forth.

Generic designs are often provided in the form of a leading-edge prototype office product, but they need not always be an actual product or prototype. They can be conceptual and yet have an important influence on user-interface design. The conceptual WIMP interface, for example (Windows, Icons, Mouse, Pull-down/Pop-up menus), has influenced the design of many products. Readers familiar with modern office systems will see the strong influence of WIMP concepts on such products as the Xerox Star, Apple Lisa and Macintosh, GEM, Tapestry, Taxi, and other products.

Through standardization activities

The similarities one can observe between many different products on the market often represent a kind of *de facto* standardization. Users learn what to expect from products in a given category because previously such products have chosen to incorporate particular features in particular ways – even if the result has been sub-optimal from a psychological point of view.

An example of what is meant by a psychologically sub-optimal solution in this context is provided in the hardware domain by the QWERTY keyboard, which has become a standard that would be very difficult to displace despite the fact that it was designed to meet non-psychological criteria and despite the fact that ergonomically better designs have been proposed. The same kind of problem will arise in the software domain unless cognitive psychologists and other human factors specialists work energetically to propose standards based on adequate research before *de facto* standards become so well established that they are as difficult to displace as the QWERTY keyboard.

Psychologists are represented on some of the key standardization bodies operating in industry but it must be said that they are currently hampered in their work by the paucity of directly relevant and applicable research. This is an area where the psychological research community must recognize that product developments will not slow down to await the results of psychological research – indeed, the product development cycle has been shortening – and if psychologists wish to influence such developments they need to develop methods of research that are suitable for producing applicable findings within the timescales that apply. Recently, a Europe-wide programme of research has been launched which addresses this problem specifically. Under the COST programme, attention is being given specifically to producing standards for user-interface design within an industry-led framework which involves a number of European academic institutions.

Through the provision of design aids

Guidelines, generic design concepts and standards help to provide the designer with a framework within which to work. However, an infinite number of possibilities exist within that framework and designers increasingly turn to psychologists for particular advice on the human factors aspects of the particular products on which they are working. In some cases it is appropriate for the psychologist to become a member of the design team or to provide advice to the team on a consultancy basis. However, there are not enough psychologists in industry to be able to cover all product developments adequately in this way. What is needed is something that can supplement these methods by providing the designer with a means of obtaining human factors feedback on particular designs other than through the direct involvement of a psychologist – at least for some of the more routine aspects.

In the older field of workstation ergonomics, an approach that has proved useful to design teams has been to provide a 'black box' that can accept descriptions of proposed layouts and provide feedback concerning their appropriateness in terms of ergonomic considerations (see Pulat and Grice, 1985, for several examples of such design aids). Even where these require a trained ergonomist to interpret the results, they can be useful in automating some of the routine aspects of the work, thereby increasing productivity and releasing some of the ergonomist's time for other, higher-level work.

Less work of this sort has been done in relation to the cognitive psychological aspects of the user-interface to office systems. However, a cooperative project led by the Human Factors Technology Centre at ITT Europe ESC-RC, as part of its contribution to the European Strategic Programme of Research and Development in Information Technology, has been working in this area (Gardiner and Christie, 1985). The other industry partners involved in the project are GEC in the UK and Logos Progetti in Italy, with contributions from researchers in several academic institutions including (in alphabetical order): the City of London Polytechnic, Manchester University, the MRC Applied Psychology Unit, Southampton University, Sunderland Polytechnic and York University.

Through user tests

User tests are important in a number of ways: for providing data on which to base design guidelines; in helping to validate and extend the principles embodied in design aids of the type discussed above; and in providing data on which to base design decisions concerning enhancements to early product prototypes.

The use of such tests can be illustrated by considering the case of text editing. One of the tasks which the user must perform is to locate incorrect characters in the text and replace them. This is usually done using cursor keys or a mouse (an example of the *de facto* standardization that is emerging in relation to office systems). Other possibilities exist in principle, and the designer must choose which to use. User tests can help provide a rational basis for the decision. Haller, Mutschler and Voss (1984) compared a number of such possibilities: the light pen, graphic tablet, mouse, tracking ball (also called bowling ball or cat), cursor keys, and speech recognizer.

The first comparison was in terms of speed of positioning the cursor over the erroneous character. In their tests, the speed with which the user could do this corresponded roughly to the degree of compatibility between the positioning method used and the system's response. The light pen was the fastest method, and speech was the slowest. They also considered how, having located an incorrect character, a user might replace it with the correct one. They compared keying in the correct character with speaking it. In terms of this criterion (speed of correcting the character having already located it) there was no difference between the two methods compared.

This example illustrates how user tests can be useful in providing data which may help a designer make a rational choice between alternative design possibilities – in this case, alternative methods for locating and correcting an incorrect character in text. The example also illustrates how the results of the comparisons can depend upon the precise combination of task and performance criterion chosen.

Very often in practice it is important to consider a wide range of tasks and a number of different evaluation criteria. This is where videorecording of user–system interaction during such tests can be of great value. A number of systems exist which allow electronic analysis of videotaped records. For example, one such hardware/software system, called CAVAS (patent pending), was developed by Christie, Argue and Haque at ITT Europe's ESC-RC Human Factors Technology Centre for the specific purpose of analysing videotapes collected for the study of office functions and user–system interaction.

The CAVAS system is based on the 'behavioural elements' approach to analysis of videorecordings. The behavioural elements approach allows the human factors specialist to break down the behaviour of interest into a number of significant 'building blocks', with a pre-defined start and finish. The classification system that gives the expert the definitions of these behavioural building blocks is compiled a priori on the basis of available research and theory. CAVAS then enables a judge, or group of judges, to view the videotape and identify the occurrence of the events of interest, which are then named and automatically tagged to specific frames in the videotape (25 frames a second) for later detailed analysis.

Through training in human factors

One of the most valuable services a psychologist can perform in relation to office systems design is to make relevant concepts and methods available more widely through appropriate training programmes for the non-psychologist members of design teams. We mentioned earlier that design guidelines can provide a vehicle for transferring human factors knowledge to the designer, though they are often used more as reference texts, or 'checklists'. Many organizations also encourage their human factors experts to prepare and teach formal training courses which are aimed at raising designers' awareness of the different types of problem involved in 'good' interface design.

Design aids such as that described earlier (Gardiner and Christie, 1985) may have a part to play in this. They provide a structured environment within which designers can be shown formal specification techniques, can see the effects of different modifications to a particular design, and can have their attention called to the reasoning behind particular evaluations.

CONCLUSION

We have considered how the psychologist can make important contributions in regard to the design of office systems in three key areas: analysis of functional requirements; development of models of user behaviour; and design and evaluation. There has been increasing awareness in industry of the potential value of psychological contributions in all of these areas. Companies such as ITT have been playing an increasing role in supporting the work of psychologists in this area by actively recruiting suitable psychologists into their own ranks, through collaborative research with the universities and polytechnics, and by providing training opportunities (for example through undergraduate industry placements and postgraduate support through the Research Councils' Collaborative Awards). The Commission of the European Communities has made funds available at the European level under its ESPRIT programme, and in the UK the Alvey programme has been very supportive of academic psychologists wishing to do research in this area.

One could argue that more could be done. US industry has generally been more supportive of the US academic institutions than UK industry has been of UK academia. IBM's support to Carnegie-Mellon University in helping to develop an 'electronic campus' has provided one model of how funds from industry can facilitate both research and education in information technology. One could argue that UK industry could provide more funds. More could be done under the national and international research programmes. The British Government could be more encouraging by providing tax incentives to industry wishing to supply the academic research communities with modern equipment suitable for the kinds of research that need to be done.

More could be done along these various lines, but it must be acknowledged that if the potential contribution of psychology is to be maximized, the onus is now on psychologists to adapt their methods of working to meet the changing requirements of this highly dynamic field – and in particular to develop methods of research and application that can provide usable results within the time-scales that are dictated by the industry. To fail to adapt to this challenge would mean that the gap between technology and psychological research would continue to widen as psychological research falls further and further behind the rapid advances being made in the technology. If psychologists can adapt, however, then not only will they have something useful to say now, but their contributions will become even more important in the future as relevant and applicable psychological knowledge about user–system interaction accumulates hand in hand with developments in technology.

REFERENCES

Argyle, M., Furnham, A. and Graham, J. (1981) *Social Situations*. Cambridge: Cambridge University Press.

Barber, G. (1983) Supporting organizational problem-solving with a work station. *ACM Transactions on Office Information Systems, 1(1)*, 45–67.

Card, S.K., Moran, T.P. and Newell, A. (1983) *The Psychology of Human–Computer Interaction*. London/New Jersey: Lawrence Erlbaum Associates.

Cole, I., Lansdale, M. and Christie, B. (1985) Dialogue design guidelines. In B. Christie (ed.) *Human Factors of Information Technology in the Office*. Chichester: Wiley.

von Cranach, M., Kalbermatten, U., Indermuhle, K. and Gugler, B. (1982) *Goal Directed Actions*. London: Academic Press.

Doswell, A. (1983) *Office Automation*. Chichester: Wiley.

Engel, G.H., Groppuso, J., Lowenstein, R.A. and Traub, W.G. (1979) An office communication system. *IBM System Journal, 18(3)*, 402–431.

Fähnrich, K.P. and Ziegler, J. (1984) Workstations using direct manipulation as interaction mode – aspects of design, application and evaluation. *INTERACT '84: First IFIP Conference on 'Human–Computer Interaction'*. Amsterdam: Elsevier Science Publishers.

Field, J. (1985) Identifying future office systems. In B. Christie (ed.) *Human Factors of Information Technology in the Office*. Chichester/New York: John Wiley and Sons Ltd.

Gale, A. (1973) The psychophysiology of individual differences: Studies of extraversion and the EEG. In P. Kline (ed.) *New Approaches in Psychological Measurement*. Chichester/New York: John Wiley and Sons Ltd.

Gale, A. and Edwards, J.A. (1985) Individual differences. In M.G.H. Coles, E. Donchin and S.W. Porges (eds) *Psychophysiology: Systems, Processes, and Applications*. New York: Guilford.

Galitz, W.O. (1980) *Human Factors in Office Automation*. Atlanta, Ga: Life Management Assoc. Inc.

Gardiner, M.M. and Christie, B. (1985) Packaging cognitive psychology for user-interface design. *Proceedings of the Ninth Congress of the International Ergonomics Association*. London: Taylor and Francis.

Grandjean, E. and Vigliani, E. (eds) (1982) *Ergonomic Aspects of Visual Display Terminals*. London: Taylor and Francis.

Haller, R., Mutschler, H. and Voss, M. (1984) Comparison of input devices for correction of typing errors in office systems. *INTERACT '84: First IFIP Conference on Human–Computer Interaction*. Amsterdam: Elsevier Science Publishers.

Kidd, A.L. (1982) *Man–Machine Dialogue Design*. British Telecom Research Laboratories, research study number 1. Martlesham Consultancy Services, British Telecom Research Laboratories, Martlesham Heath, Ipswich.

Lum, V.Y., Choy, D.M. and Shu, N.C. (1982) OPAS: An office procedure and automation system. *IBM System Journal, 21(3)*, 327–350.

Marshall, C., Nelson, C. and Gardiner, M.M. (in press) Design guidelines. In M.M. Gardiner and B. Christie (eds) *Applying Cognitive Psychology to User-Interface Design*. Chichester/New York: John Wiley and Sons Ltd.

Murray, D. and Bevan, N. (1984) The social psychology of computer conversations. *INTERACT '84: First IFIP Conference on Human–Computer Interaction*. Amsterdam: Elsevier Science Publishers.

Pulat, B.M. and Grice, A.E. (1985) Computer aided techniques for crew station design: Work-space Organizer – WORG; Workstation Layout Generator – WOLAG. *International Journal of Man–Machine Studies, 23*, 443–457.

Richards, M.A. and Underwood, K.M. (1984) How should people and computers speak to each other? *INTERACT '84: First IFIP Conference on Human–Computer Interaction*. Amsterdam: Elsevier Science Publishers.

Smith, S.L. (1982) *User–System Interface Design for Computer-Based Information Systems*. Reference ESD-TR-82-132, The MITRE Corporation, Bedford, Massachussetts.

Van de Ven, A.H. and Ferry, D.L. (1980) *Measuring and Assessing Organizations*. Chichester/New York: John Wiley and Sons Ltd.

Part 2
IT IN HEALTH AND EDUCATION

Increasingly, information technology is being used to the benefit of members of society in general, in addition to that of specific groups or organizations. These social services of the technology can be seen particularly in the health and education fields, in which the technology's ability to store and retrieve very large amounts of information with speed and efficiency enables records to be kept and data bases to be interrogated. Naturally, this leads one to consider the computer as being analogous to a very competent secretary or other type of expert. In other words, to be endowed with some level of intelligence (albeit artificial).

In the first chapter some of the reasons why computers may not be accepted in the health field are considered by Fitter. These relate primarily to the extent to which the computer may be seen as an intruder into an otherwise stable and trusting relationship between doctor and patient. Nevertheless, by understanding the function and use of such technologies, Fitter argues, lessons can be learned about how such technologies can be introduced with increasingly beneficial results. He argues that 'patient-centred' rather than 'procedure-centred' systems are needed.

Thomas considers the role of the computer as an expert in the health field, and sets the scene for the various ways in which it can be used. As the science of behaviour, psychology has a great deal to contribute in this area. By understanding how information is stored, retrieved and integrated by the expert, suitable (artificial) systems may be produced that will act as an aid to the human operators within the health care field. As Thomas points out, psychology also has a role to play in assessing the acceptability of such systems – both in terms of accuracy and human attitudes.

A specific application of the health field is considered by Hales in his chapter discussing IT applications for the disabled. As Hales points out, the benefits of IT to disabled people can be very considerable indeed. His discussion underlines the primary themes of the book: effective application of IT for people will occur only when the design of the system is matched adequately to the needs and capabilities of the user.

Finally in this section, Underwood and Underwood take the application of IT into the education field and consider its use in the classroom of the future. They extend the use of computers from applying their abilities for storing and retrieving large amounts of information to include their potential for unlimited patience and extending the pupil's capacity for exploration. In a discussion which emphasizes once again the function and usability of the technology, they point out how the psychologist's knowledge and experience in understanding the developmental processes of children can play a large part in helping to design software which matches their abilities.

THE DEVELOPMENT AND USE OF INFORMATION TECHNOLOGY IN HEALTH CARE

Mike Fitter

The application of information technology and psychology to the field of health care does not have a long or impressive past. However, it does have an important future. This book is a testament to the conjoining of information technology and psychology in an endeavour to analyse, and thereby improve, the effectiveness with which information technology can be developed and used for a variety of purposes. However, as yet there have been relatively few applications of information technology in the provision of health care, and extremely few that have had the active involvement of psychologists or used the techniques of psychological research. Several past attempts to develop computer systems for aiding the provision of health care have been far from successful, and could have benefited from an input from psychology, particularly from organizational psychology.

This chapter describes and assesses some recent applications of information technology in health care, and proposes a conceptual framework based on the perspectives of principal 'stakeholders'. The second half of the chapter focuses on recent projects which have involved psychologists in the design and use of information technology in health care. It is argued that one important contribution from psychology is to *evaluate* developments, and provide *guidance* on ways of increasing effectiveness in the organization and beneficial effects for the providers and receivers of health services. From a comprehensive and systematic assessment of current projects, lessons can be learned for the future, and the replication of mistakes avoided.

One of the themes developed in this book is that the use of

technology does not *determine* the way that work will change: that, although a commonly occurring consequence of introducing new technology is to deskill work and increase managerial control over the work process, this is not necessarily the case. In fact if it were, it is very unlikely that such professional groups as doctors would support its adoption. Yet the opposite of determinism, unrestricted choice, does not exist either. In this chapter I hope to demonstrate the ways in which information technology, if used effectively, does constrain work processes and thus limits the autonomy of service providers. The aim is to articulate the circumstances in which information technology constrains choice, and the circumstances in which a full range of alternatives still exists, even though there may be *economic* pressures for certain organizational forms to emerge.

THE ADOPTION OF INFORMATION TECHNOLOGY

The adoption of information technology in health care has been limited in the past. This appears to have been for two principal reasons – there have been only limited incentives to do so, and the practical difficulties have been considerable.

Health care in the UK is mainly part of the public sector, which does not have a management tradition of maximizing 'profits', nor, until recently, has it systematically tried to reduce costs. Since conventionally one of the main objectives when introducing information technology has been as part of a strategy to reduce operating costs (at the expense of increased capital investment), it has not been regarded as a priority in health care. However, the indications are that this perspective is changing rapidly. For example, within the public sector generally, the social security computerization strategy was set up to reduce employment in the service by some 20 000 to 25 000 jobs over a 10-year period, at a capital cost of some £700 million. Whether or not these plans are realistic remains to be seen, but the indications are that similar strategies will develop within the health service. The much publicized 'Griffith's proposals' (*The Times*, 26 October 1983) include a statement that each hospital should develop management budgets which invoice doctors and relate workload and service objectives to financial and manpower allocations, and that managerial control should be in the hands of general managers at regional, district or unit levels. These developments have already led to cost-cutting exercises at the periphery of health care, for example in the provision of domestic services.

Information technology may play a significant role in further rationalization and efficiency promotion strategies. However, the process of rationalization at the core of health care would not be straightforward, for the following reasons:

- The medical profession still has a substantial role in health-care planning and decision making. By and large it neither has much experience in management rationalization techniques, nor is it likely to show much enthusiasm for changes which could result in less autonomous working conditions.

- Using information technology to rationalize the delivery of care may lead to specific reductions in the quality of the service provided to patients, and may lead to depersonalizing of health care. Such considerations are vital to personal service organizations, and are in contrast to activities in manufacturing industry, where the output from a production process is an inanimate object. The tradition of the medical profession, which extends back to the Hippocratic oath, is to put first the interests of the individual patient. Thus considerations by which 'cost effectiveness' determines priorities for resource allocation have second place.

The application of 'value for money' accounting would require attention to resource allocation and an explicit and 'rational' assessment of priorities and competing demands. Such management techniques have been applied within the hospital sector, and an example from a project in a hospital in the 1970s illustrates some of the above points.

A computer team at Addenbrooke's hospital in Cambridge was concerned with the less than optimum way that hospital resources were being used. One particular case was the utilization of the operating theatres (Hammersley, 1975). The normal procedure was for the surgeon responsible for an operation to book the theatre for as long as it was anticipated would be necessary. Analysis of data revealed that an operation booked for four hours took on average 192 minutes with a standard deviation of 90 minutes. Using the techniques of operations research the team discovered that, based on the characteristics of individual surgeons and the type of operation to be performed, it could predict more accurately than the surgeon how long an individual operation would actually take. Since more accurate predictions would save money (more operations carried out in the facilities available, and less overtime payments to technical staff, paid when operations overran) it was proposed that these management techniques be used in the allocation of resources. Perhaps not surprisingly the surgeons objected, arguing that the decision was necessarily one of 'clinical judgement', requiring that many additional factors be considered when allocating theatre time. In particular they were concerned about special cases such as teaching sessions. In the event the proposed system was not implemented.

The failure of this initiative indicates some of the complexities and problems likely to be encountered. Note that the problems are not 'technical', nor concerned with the use of information technology *per se*, but stem from the concerns of interested parties about the consequences of organizational change. They are part of a political process.

Yet, in case the above analysis conveys the view that health care providers are adverse to technology, it is important to point out that the development of modern medicine is almost synonymous with the development of medical techniques and technologies. The 'scientific' approach to medicine, sometimes referred to as the 'medical model' of health care, develops and applies techniques of medical cure. The more popular areas of health care within the medical profession, and those prominent in the public eye, are the high technology areas. For example, reports of organ transplants regularly appear in the media, and fund raising for body scanners is commonly achieved by public donation. The technology and its application have a glamour for the public and the health care providers alike. It is able to attract a substantial proportion of the health service's resources. This is of particular concern in the private sector, which has a more explicit concern with cost-effectiveness. The chief executive of BUPA is quoted as saying recently, 'There is a need to use facilities more efficiently. We need to educate consumers as well as providers – people can be titillated by new technology' *The Observer*, 27 April 1986).

PRACTITIONER TOOLS VERSUS MANAGEMENT AIDS

The important characteristic of these applications of advanced technology is that they are tools in the hands of doctors. Their development has occurred in close collaboration with the medical profession, and their use is under the direct control of doctors. They increase the sophistication of patient care, and appear to enhance the skill requirements and status of the medical profession. Perhaps somewhat surprisingly, there has been relatively little systematic evaluation of the effectiveness of these expensive medical techniques (Council for Science and Society, 1982).

In contrast, applications developed by hospital computer services departments have been management aids, developed and installed under the guidance of health service managers. They are usually based on a formal model of the organization which simplifies the realities of day-to-day work practices. The techniques of systems analysis are essentially 'reductionist' and based on a sub-division of tasks and a standardization of work processes. They may thus be seen as a threat to the 'clinical freedom' of doctors.

This contrast between practitioner tools and management aids is, of course, not sharp. Some applications will offer opportunities to enhance both the work of the practitioners *and* provide information for a management control system. Bock (1982) has observed that many new medical systems 'piggyback' information collection, for management or research purposes, onto a system designed to directly improve patient care. Bock argues that the most successful systems will be ones that result from decisions to address real patient care and provider needs as the priority.

A further complication is the dual role held by some medical professionals. Hospital consultants, in addition to being practitioners, have management responsibilities on hospital committees. Child (1984) suggests that through this managerial experience:

> These higher level workers are often instrumental in initiating the rationalisation process, which may start off as an attempt to cope with perceived overload and complexity. They may previously have established a precedent by assisting in the rationalisation of lower-level white collar work (p.1).

In the UK general medical practitioners (GPs) would first and foremost identify themselves as practitioners. But they are also the owners of small businesses, and the employers of ancillary staff. Thus in the development of information technology for general practice, GPs have a particularly ambiguous role. The use of information systems in the primary health care sector is seen as particularly difficult, and there have been relatively few developments to date. Davies (1984) comments:

> the community health services have become a management 'no-go' area. The new information systems must be designed to support local management in a systematic approach to decision-making (p. 189).

But what will practitioner tools or management aids offer for the patients? Do they, and will they, result in a better standard of service which patients will experience more positively? There has been little research which addresses these questions. Assessments that have been made of new techniques usually focus on clinical efficacy – that is effectiveness within the restrictive 'medical model'. Few studies have been carried out from the patients' perspective, although this is a topic well suited to the techniques of psychological research.

THE STAKEHOLDERS IN THE PROVISION OF HEALTH CARE SERVICES

A vital element in any framework for the development and assessment of information technology in the provision of personal services is an identification of 'stakeholders' who have an interest in the changes that will occur. It is important to recognize that the stakeholders will have their own perspective on the service in question and the proposed developments which will change it. They will have their own goals and priorities which may or may not be compatible with those of other stakeholders. The stakeholders and their interrelationships are represented in the triangle.

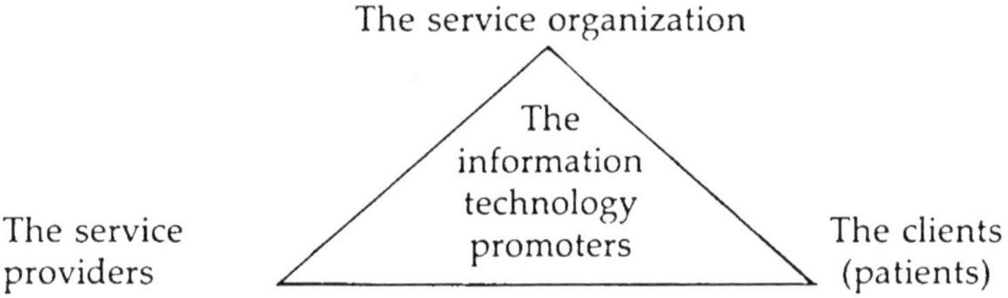

There are three principal groups of stakeholders involved in the provision of health services, and these are all likely to be influenced by changes initiated by the installation of information systems. A fourth group has a specific and primary interest in promotion of the technology. A brief description of the perspective of each group and its relationship to the other groups follows.

The service providers

Professional groups are the 'front line' in the provision of health services – doctors, nurses, physiotherapists, psychologists, etc. The core characteristics of the professions, and of the medical profession in particular, have been identified as:

- a high degree of autonomy, for example, the profession determines its own standards of education and training, and is relatively free from lay evaluation and control
- a service orientation; a commitment to 'look after' the needs and interests of the client group (Friedson, 1970).

Service providers have an interest in maintaining or enhancing their skills and status. Thus the development of new techniques or technologies, if they are to be acceptable to professional groups, should aim to improve the service to patients *and* maintain the autonomy and work quality of the service providers.

It is argued here that the degree to which service providers accept new information technology will depend on:

- their perceptions of its effectiveness in improving the service to clients
- their perceptions of their ability to use and benefit from technology, and in particular their perceptions of its effect on autonomy, job content and skill requirements
- their perceptions of the need to *change* the current work situation
- their personal investment (financial or psychological) in the developments.

The service organization

The greatest part of health care in the UK is provided by the National Health Service, yet within this a wide range of organizations exists. For example, at the primary care level there are some 10 000 general practices, grouped into districts and administered by Family Practitioner Committees. At the secondary care level there is a wide range of hospitals and specialist units of greatly varying size. Within the conceptual framework developed here, the organizations, represented by senior management, have the common purpose of administering the provision of health services. They also have an increasingly explicit goal to *manage* the services. This enhanced goal is to set organizational objectives, monitor activities and exercise *control* over the operation of the organization. Cost-effective use of resources and reduced labour costs per unit of service is the concrete aim.

One important factor when introducing information technology is to consider how it will affect the balance between management control over resources and the autonomy of the service providers.

The patients

We are all patients, though for most of us, it is a part-time role. Perhaps for these reasons it is difficult to make generalizations about the characteristics, needs and interests of this group of stakeholders. It is also the case that patients, as an interest group, would not usually have any direct input into decisions over the development and use of information technology in health care. Thus, compared with the previous two groups, their interests cannot easily receive direct representation. Both health care providers and health service organizations would claim to include the interests of patients under their own interests. It is nevertheless worth examining what perspective patients might have on the development of information technology, and how technological change might influence their relationship with the other stakeholders. This perspective is likely to become more prominent in the future as a more 'consumer' oriented view of health care is promoted (see, for example, the recent Government Green Paper *Primary health care: An agenda for discussion*, 1986).

The first concern of patients is that they should receive a high standard of health care, readily available at reasonable cost. The emphasis on availability and cost will be influenced, in particular, by whether a 'market' relationship exists between the client and the service. For a service provided by the welfare state (such as health care), availability will be more of a concern than cost. For a service provided through the market place (such as solicitors' services) cost is likely to be of as much concern as availability, unless prices are fixed. Another concern, at least for some, is that they should be able to have a confidential and personal relationship with their doctors. Thus surveys of patients' attitudes to the use of computers in health care have revealed some concerns about a threat to confidentiality of records and a potential loss of 'the personal touch' (Cruickshank, 1984; Pringle, Robins and Brown, 1984).

Another, more contentious, concern is that patients should play a more active role in their health care. A characteristic of the patient's role is that it is dependent upon, and subservient to, the superior knowledge of the doctor. In relation to the patient, the doctor's role has been described as authoritarian in approach (Tate, 1983). Will the use of information technology change this relationship? Will it provide the patients with the knowledge required to participate as an equal in the doctor–patient communication? Or, conversely, will the use of computers in the consultation consolidate the doctor's position as expert? The outcome will be a change in social relations, and possibly in clinical effectiveness. Fitton and Acheson (1979) have stated 'there is little doubt that a relationship based on equality requires not only more giving on the part of the doctor, but his relinquishment of superior power in the dyad.' Tate (1983) has also observed that this would 'probably decrease some of the doctor's authority and, although understanding in the patient is increased, perhaps the magical healing and anxiety relieving power is actually reduced' (p.77). Only some patients appear to want a more active role and a less dependent relationship with their health care providers. Nevertheless, as will be elaborated, the relationship is likely to be affected by the way that information technology is developed and applied.

The information technology promoters

The development of information technology in an organization may be initially promoted by one of the three groups of stakeholders directly involved in the service (though the initiative is unlikely to be from the clients). However, except possibly in a very small scale development, professional systems developers, with their own 'technical' perspective, will soon become involved. This group of stakeholders includes system suppliers, analysts and trainers from outside the organization, and in large service organizations may include specialists such as are found in data processing departments.

Even after the initiation of a new development, providers such as doctors may maintain a keen interest and close involvement. In these situations it is not uncommon for doctors to adopt, at least temporarily, a technical perspective. For example, in our experience, some GPs become so enthusiastic about the processing power of their new micro and the potential of the new 'expert system', that they appear to overlook the day-to-day realities of patient care!

In situations where professional system developers are used, it is important that their designer perspective does not dominate. The legitimacy of the service provider's view of the organization must recognized, even though it may be only implicit, and therefore need to be articulated by a sensitive development process. Nygaard and Sorgaard (1985) have warned:

> The system analyst exercises perspective power and establishes a perspective monopoly by insisting upon and succeeding in the exclusive use (in the development process) of facts, experience, concepts, techniques, and tools that are meaningful within the framework of a system perspective (p. 2).

A MULTI-PERSPECTIVE APPROACH TO INFORMATION TECHNOLOGY DEVELOPMENT AND USE

The initiation of a new development, and the forces which steer its progress, will be influenced by the above perspectives. Clearly each stakeholder group will itself have multiple perspectives and interests, which it will share with other groups. There will also be differences within each group. For example, doctors and nurses are both service providers, but differ in their perspective on health care. National and local health service organizations may also have differing perspectives. Yet each group has its own base perspective which, in the development and use of information technology, can be specified as (in reverse order):

— a technology-led approach: try it and see
— an improvement in the quality of the service
— an increase in the cost effectiveness of the service
— an improvement in the quality and conditions of work.

These perspectives will influence, and substantially determine, the outcome of developments. Thus the introduction of information technology into health care is not primarily a technical process, it is a political process bringing about new social relations. From *this* perspective it should not surprise the reader that projects that are handled solely as 'technical' developments, promoted by technologists from a technological perspective, will fail. What does seem surprising is that major developments *are* commonly managed from a technological perspective and commonly *do* fail, or at least fail to live up to their potential. The rest of this chapter describes some of the ways in which psychological research has contributed in the recent past to the development of information technology in health care, and some of the ways it could contribute in the future.

EMPIRICAL STUDIES OF THE APPLICATION OF INFORMATION TECHNOLOGY TO HEALTH CARE

This section reviews some of the research that has been reported recently on the use of information technology in health care. It focuses on the contribution of psychological research to field studies of technology in day-to-day use. It is in this way that the issues and priorities identified in the previous section can be pursued most directly. Within the framework of stakeholder perspectives that has been established, this section uses examples from primary health care to examine the influence of information technology on the service providers and their relationship with the patients.

Information technology in the direct provision of health care

The quality of the relationship between health care providers and their patients is generally regarded as important and in need of preserving. Concerns have been expressed that information technology may, somehow, get 'in between' both participants if used in the direct provision of health care. For example, a report from the Royal College of General Practitioners (1980), whilst welcoming the introduction of computers, cautioned:

> We have one important reservation about this development. We do not know whether direct input to the computer during the consultation will have an effect on doctor/patient communication. Research on this problem is urgently required (p. 9).

Since that time research has addressed this question, and the results are reviewed here.

Computers in the medical consultation. Several studies have been carried out by researchers in the Social and Applied Psychology Unit at Sheffield University on the use of computers in the consultation. They all involve the use of terminals on the doctor's desk. The systems were designed to assist with a range of applications, including the history taking and diagnosis of patients with symptoms of dyspepsia, medical record keeping in the general consultation, the management of patients with hypertension and use by midwives for the collection of ante-natal histories. The general methodology of all the studies has been similar. Consultations involving the use of the computers have been compared with 'ordinary' consultations. The studies have examined the impact on the provider's work, the consultation process and the patient's perceptions of and satisfaction with the consultation. This has been achieved by an analysis of medical and computer records, of video-recordings of consultations, and of interviews and questionnaires which ascertained the views of the providers and patients. Care has been taken to ensure that ethical guidelines for the pursuance of research have been adhered to.

Within the confines of this chapter it is only possible to summarize the results and sketch out the picture that is emerging. The interested reader is invited to pursue the findings in more detail via the references provided.

Structuring of tasks. A design choice exists as to how much the user should be constrained by the information system. There are two principal ways in which the user's choice may be limited. Firstly, the computer may impose *structure* on the consultation process. For example, a protocol for the management of chronic conditions specifies information which should be obtained from the patient and decisions which should be made, based upon that information. Results on the use of a hypertension protocol indicate that, provided the user approves of the protocol and has the option to override the computer, this degree of structuring is acceptable.

History taking and management protocol systems increase the likelihood that the user will collect information directly relating to the system, and this may be at the expense of discussing other topics with the patient (Brownbridge *et al.*, 1986; Pringle, 1985). Thus the computer tends to focus the consultation, which may or may not be appropriate. There is also evidence that because the computer 'invites' specific information, specific symptoms are likely to be entered even when they have not been explicity elicited from the patient (Brownbridge, Fitter and Sime, 1984; Brownbridge *et al.*, 1986). A history taking system need not structure the sequence of information gathering if the system is designed to provide user flexibility (Fitter and Cruickshank, 1983). However, a management protocol must necessarily structure and thereby constrain the consultation process because that is its purpose.

The second way in which a computer may impose structure on the consultation process is by requiring information to be categorized in a predetermined way. A history taking system or diagnostic aid requires the user to adopt and adhere to specific definitions of symptoms, etc. Normally doctors are free to choose their own classification system – it's up to them what they write in the patient's notes. For a computerized system to be effective standardized definitions need to be used. These must be discussed and agreed by the doctors in a general practice partnership (assuming they see each others' patients, as is common) and possibly more widely within the medical profession. If the computer were to provide a *tabula rasa* to the doctor, an unstructured screen able to take 'free-text' akin to the traditional patient notes, the doctor's record keeping would not be constrained. However, the main *benefits* of computerization result directly from operating within an agreed structure, and using a pre-determined categorization system. Without a structure, information is extremely difficult to retrieve or analyse for decision making purposes (Evans and Brownbridge, 1985; Fitter, 1986*a*). Yet the imposition of an information structure (voluntary or otherwise) constrains the doctor's 'clinical freedom' in ways that may be unacceptable to some, and requires that doctors receive *training* in the use of the information structure.

In a similar vein, Dewey and Copeland (in press) argue that a benefit of using a computer for the classification of symptoms used in the psychiatric diagnosis of the elderly is that 'computer methods . . . enforce a precise statement of the theory involved'.

To summarize, computerization of medical information offers more consistent, legible, retrievable and analysable products. But it also requires a higher level of agreement between the service providers, more training to produce the necessary level of competence in the use of standardized information procedures and a commitment from the users to adhere to those procedures.

Clinical effectiveness, operational efficiency and opportunity value. An important aspect of an information system is its task-related effectiveness. In our assessments of experimental prototypes in the consulting room a fairly consistent pattern has emerged. The systems have, by and large, had little direct impact on clinical performance. The ability of registrars to diagnose dyspepsia symptoms at the patient's first visit to an out-patient clinic was similar with and without computer assistance (Fitter *et al.*, 1983). In general practice consultations, peer group assessment of the GP's clinical performance revealed no significant overall difference between computer and non-computer consultations, although there was a tendency for the computer consultations to be rated lower on several of the performance measures (Brownbridge, Evans and Wall, 1985). However, the use of the hypertension protocol by GPs did result in an improvement in clinical standards of health care delivery, although no assessment of health outcome was possible (Brownbridge *et al.*, 1986).

In terms of operational efficiency, the measure most relevant to the doctors conducting the consultations was whether the computer extended or shortened the time required to process information. In all the studies, the use of a consulting room terminal has lengthened the time spent on information processing. It would be unlikely to be otherwise, because analysis of pre-computer consultations revealed that very little time was spent on information recording or retrieval

(Fitter, Garber and Wall, 1982). The computer may be able to improve the standard of information keeping, but it is unlikely to shorten the process. It is interesting that system promoters claimed that the computer would save time on record keeping. Yet a relatively small investigation of GPs' *actual* record keeping would have revealed this was very unlikely in practice.

The third aspect of system effectiveness is the extent to which it has 'opportunity value'. Although the computer may not save any time during the consultation, it may provide the means to improve the quality and comprehensiveness of information recording. This may in the longer term lead to more systematic processing of information, more *appropriate* retrieval of information, more audit of information and ultimately better standards of medical care. The general pattern that emerges from our studies is that in the short term most of the computer effects are costs to the user; the benefits (many of which have yet to be realized) appear to lie in the longer term. Thus, for use in the direct provision of health care, information technology should be regarded as an investment with a long term return, rather than a short term gain. This view would offer users a realistic perspective, but is not necessarily compatible with the marketing strategies of information technology promoters.

The effect on service providers and patients. The doctors who have used consulting room systems have found the experience mildly stressful. This appears to be largely due to the time that computer use takes, but may also be a consequence of their relative inexperience and poor typing skills. In particular, doctors who characterized their consulting style as 'non-directive' had more negative views and thought the computer might threaten their relationship with the patients. In contrast, doctors who saw themselves as 'directive' thought that the computer could be a useful tool in managing the flow of the consultation. These were the doctors who used the computer most (Herzmark *et al.*, 1984).

Detailed analysis of the video recordings revealed only minimal effects of computer use on the doctor–patient communication process. Analysis of verbal communication provided evidence that doctors compensated for the computer by showing more solidarity with patients, through affective statements and by injecting humour (Brownbridge, 1986). Assessment of patients' attitudes to the use of computers by doctors indicated that patients who had direct experience of the doctor using a computer had more positive attitudes than ones who had experienced an 'ordinary' consultation (Fitter and Cruickshank, 1982).

Overall, the indications are that, although care should be taken about how computers are used, particularly in sensitive consultations, the expressed concern that computers will depersonalize the doctor–patient relationship is relatively unfounded. The evidence suggests that a 'computer technique' can coexist alongside some of the other more 'human-centred' techniques (Fitter, 1986*a*).

The integration and control of health care. The conventional role of the doctor is to provide an acute service – to respond to patient-initiated requests. However, there is a growing belief that this approach is too piecemeal and results in fragmented health care. In particular, it has been suggested that GPs should adopt a more proactive role that aims to provide an integrated service. It has been argued that there should be a shift towards more anticipatory, or preventive, health care (Marinker, 1984). This shift in emphasis can be assisted by the development of information systems in primary health care – as is already beginning to happen for preventive care programmes; for example in cervical cytology screening.

However, if prevention is seen as a set of 'medical procedures' (tests, immunizations, etc.), as many GPs tend to regard it, then instead of developing an integrated service it is likely that a more fragmented one will result (Fitter, 1986*a*). Using computers in the management of preventative procedures enables GPs to take a more proactive role (calling patients to the surgery at predetermined points, etc.), but it may lead to patients developing an even more passive role, since they will have *less* need to initiate a request for medical attention.

Stott (1985) has argued that the technology *can* be used to provide more integrated health care:

> The hardware now exists for machines which can enhance the power of consultations by integrating epidemiological and clinical information for each patient's benefit. Clinicians will then be able to act more effectively and efficiently in every consultation, teamwork will flow more logically and the patient (client) will know that when a visit is made to the health centre he or she will receive much, much more than episodic care. The challenge for the future is first for clinicians to learn to approach each consultation in a broader way, and secondly for the systems analysts to provide the software which will encourage this integrated approach (p.142).

General practice systems now exist which can provide the doctor with 'opportunistic prevention' prompts during a 'routine' consultation. The GP has a profile of the patient which includes indicators of which topics are 'due'. For example, the computer may prompt the doctor to invite the patient to have a tetanus booster, or to enquire about the patient's smoking habits (Shepherd, 1984). The advantages of opportunistic screening are twofold. First, patients do not need to be written to (they are there anyway – the majority of patients visit their doctor at least once during a 12-month period), and therefore a more personal and educational approach to the subject of prevention is possible. Second, instead of each preventative procedure being considered separately (and possibly being provided by a paramedical member of staff without the direct involvement of the doctor), the doctor can make an *integrated* assessment of the various requirements, perhaps deciding to defer suggesting a particular test or immunization because of the individual circumstances.

The use of information technology by patients. In order to participate more actively in their own health care people need an adequate understanding of the causes of health and ill health. There is a tendency for 'health' to be seen passively as the absence of illness. Health education is a vital aspect of health care which can change this view. One possible application of information technology is in the provision of health education information. Relatively few developments have taken place in this area, although software is beginning to emerge, designed primarily for the consumer market of people with home computers. Examples of such software include *Beat the Bottle* (Hopley, 1986), an aid to reducing alcohol consumption, and *How Long Have You Got?* (Coleman, 1983), a health education aid.

These aids are designed for independent use by members of the public. They may recommend seeing a doctor in specific circumstances but the decision is up to the user. An alternative approach, which has more similarity to the history taking aids described earlier, is illustrated by a growing number of systems which elicit histories, or other information, from patients. The information is then used by the health care provider (usually a doctor) for assessment. The aim of such systems is typically to increase the reliability of symptom elicitation (by reducing observer error) and to reduce costs (by saving the doctor's time, even if it increases the time the patient spends) (Knill-Jones, Dunwoodie and Crean, 1985). Studies have shown that computer interrogation, as it is often called, is generally liked by most patients (Lucas, 1977) and may result in more accurate information being elicited (Lucas *et al.*, 1977), although this latter finding was not replicated by Skinner and Allen (1983).

Although the indications are that these systems can be effective if well designed, they are likely to reinforce the passive role of patients as part of a 'production process'. The situation would be different, of course, if the output from the system were provided for the patient, who then decided how to proceed. This is generally not the case. However, such 'educational' information *could* be provided, either in the consulting room by the appropriate redesign of doctor-centred systems, or through the further development of systems for use outside a medical environment.

Information technology in the management of health care

Computers offer the opportunity for more effective managerial control. The extent to which GPs want this in their own practices is unclear. It has been argued that the main benefit to GPs of computerization is a cost-effective opportunity to feel more in control of their practice (Herzmark, 1986). Audit of practice activities is necessarily part of a management system and can make an important contribution to making computer use effective (Fitter, Evans and Garber, 1985). Yet GPs may be reluctant to open themselves to such scrutiny, even if it is confined to peer group review:

> Most of us face some difficulty with the word 'audit' and its implication of outside supervision and loss of 'independence'. The status of 'independent contractor' has in general practice become such an article of faith that audit is seen to be sacrilegious. This is compounded by the professional isolation of most general practitioners who, in the main unsupported by colleagues or young doctors in training, live in the small world of their consulting rooms. Unused to either mutual support or scrutiny, we become more and more paranoid about outside interference (Drury, 1981, p.259).

One of the most comprehensively evaluated recent developments in GP computing was the Micros for GPs scheme which, initiated by the Department of Trade and Industry in 1982, provided subsidised microcomputer systems to 150 general practices. The systems provided facilities to support the administration of practices. One of the main conclusions from this scheme was that:

> There is a growing awareness that the main benefits of computers do not lie in the mimicry of manual procedures for the day-to-day routine of general practice administration. Rather it is

> in the aggregation and analysis of information that the computer's analytic strengths seem to offer real advantages (DHSS, 1985, summary).

The report goes on to argue that this requires a systematic approach to using information, and in essence the computer is really a *catalyst* for change. This is illustrated by a quote in the report from one of the GPs' annual progress reports:

> When the computer is looked at from the doctors' point of view in our practice I regard it as an expensive catalyst. It has facilitated changes in the practice which, given drive and determination, could quite easily have been achieved without its presence and at less cost. The computer has provided two things for the secretaries, taking over the chore of writing repeat prescriptions and secondly providing employment for another half-time secretary (p.86).

This illustrates another conclusion in the report, that, contrary to popular belief, computerization *increased* the need for administrative staff. Overall, the report found little evidence of substantial concrete benefits arising from the introduction of computers, and probably insufficient to justify the expenditure incurred. Yet most of the doctors (whose money was involved) and their staff felt enthusiastic about computerization after the first year of use.

Herzmark (1986) suggests that this may be due to one or more of the following factors:

(i) cognitive dissonance because of the effort involved
(ii) prior anxieties about what might happen had been allayed; for example, the receptionists had not lost their jobs, and had coped with the demands of the new tasks
(iii) general practice can become rather routine for GPs and the computer offered new opprotunities (the opportunity value)
(iv) the doctors *felt* more in control of their practice
(v) both doctors and staff thought that the introduction of computers was *inevitable* and they had so far *survived* the experience unscathed.

Explanations (ii), (iii) and (iv) are illustrated by the following quotation from the *Micros for GPs* report:

> Perhaps the most important point here is how the scheme has raised the general consciousness about the benefits of computer-

ization and taken away the fear . . . General practice is something all my partners have come into on positive grounds, i.e. on the basis of a desire to do general practice . . . Nevertheless after a few years experience everyone goes through the stage of finding at least part of the work humdrum, and tedious, and we all feel that the ability to pick an interesting subject out and study it in depth is a major advantage (DHSS, 1985, p.85).

The idea of inevitability (v) is particularly interesting because it seems to tap a popularly held belief about computerization; and yet 'objectively' the doctors, as owners and managers of their small businesses, have a 'free choice' about how their organization develops.

The introduction of computers required the acquisition of new skills in the use of the keyboards and the computer functions, but also in understanding the requirements of establishing and maintaining an effective information system. The functional aspects are usually provided adequately by the system suppliers, but the latter, which require organizational skills and techniques, are usually overlooked, and seriously impede the effectiveness of computerization (Fitter, 1986*b*).

In the Micros for GPs scheme the main computer users were the administrative staff (receptionists and secretaries). Some practices chose to create the specialist role of computer operator (and, interestingly, at least two practices appointed a male computer operator; all receptionists and secretaries are female), others chose to spread the computer tasks across the existing staff (often increasing the overall staff hours as well). The specialist strategy requires fewer staff to be trained, and may therefore be organizationally cheaper to implement. However, it may also create problems. It can create an 'us and them' feeling amongst the staff, especially if the computer operator is new to the practice rather than a retrained receptionist. Analysis of the Micros for GPs scheme data revealed that, although staff who used the computer were more positive about its usefulness than those who did not, there was evidence that those who used it substantially (more than two hours per day) became less positive over 12 months of use. In contrast, receptionists who used the computer to a lesser extent (less than two hours per day) became more positive with experience (Sims and Herzmark, 1986).

It should be emphasized that the implementation strategy is clearly a managerial choice, and not a unique outcome of computerization.

SOME IMPLICATIONS

Computerization in health care presents two important strategic choices. The outcome of neither is *determined* by the use of information technology, though both need to be addressed in any major development.

First, it is possible to adopt a 'patient-centred' approach, which attempts to provide integrated health care, maintaining the quality of contact between health care providers and patients. The alternative is a fragmentation of health care, by a 'reductionist' analysis of health care into a series of *procedures* which can be applied separately and less personally (possibly by paramedical technical staff).

Second, it is possible to distribute the information technology applications to all or most of the health care workers, developing their roles and skills accordingly. Alternatively, specialist technical roles can be created, thereby increasing the division of labour and the fragmentation of health care.

These strategic choices are connected. Both concern integration versus fragmentation, though the first directly concerns the service to patients, and the second the quality of work for providers. In this respect both groups of stakeholders have the common interest of promoting the integrated approach; and the information technology, perhaps more than previous technologies, has the potential to provide it. Yet is this the direction that change will move in? The signs are not particularly hopeful. That is why it is important for psychologists to work with other professionals on developmental projects, assisting with the articulation of alternatives and the design and implementation of systems.

Systems development is traditionally seen as a *technical* process. In reality it is a *social* process that can substantially change relations between service providers, clients and the service organization. It is important to identify the perspectives of the stakeholders in a proposed development, and to articulate the likely consequences of a particular alternative for each group. It is also important to monitor and systematically evaluate developments, and to feed back results, as part of an ongoing process. Psychologists have the necessary analytic and developmental skills to work with the groups involved, provide support and facilitate change.

ACKNOWLEDGEMENTS

I would like to acknowledge the helpful comments on a draft of this chapter from Jean Hartley and from colleagues Garry Brownbridge, Bob Garber, Guy Herzmark and Toby Wall, who have also contributed to the research and ideas reported in the chapter.

REFERENCES

Bock, F.M. (1982) Considering human factors in the initial analysis and design of a medical computer system. *Journal of Medical Systems, 6,* 61–68.

Brownbridge, G. (1986) Computers in the consultation: Effects on doctor – patient communication. Paper presented to the British Psychological Society Annual Conference, Sheffield.

Brownbridge, G., Evans, A., Fitter, M. and Platts, M. (1986) An interactive protocol for the management of hypertension: Effects on the general practitioner's clinical behaviour. *Journal of the Royal College of General Practitioners, 36,* 198–202.

Brownbridge, G., Evans, A. and Wall, T. (1985) The effect of computer use in the consultation on the delivery of care. *British Medical Journal, 291,* 639–642.

Brownbridge, G., Fitter, M. and Sime, M. (1984). The doctor's use of a computer in the consulting room: An analysis. *International Journal of Man – Machine Studies, 21,* 65–90.

Child, J. (1984). New technology and the 'service class'. Paper presented to the Annual Conference of the British Sociological Association, Bradford.

Coleman, V. (1983). Home diagnosis. *Practice Computing, 2(1),* 14.

Council for Science and Society (1982). *Expensive Medical Techniques.* London: CSS.

Cruickshank, P.J. (1984). Computers in medicine: Patients' attitudes. *Journal of the Royal College of General Practitioners, 34,* 77–80.

Davies, P. (1984). Körner: The implications for community nursing. In B. Kostrewski (ed.) *Current Perspectives in Health Computing.* London: British Computer Society.

Dewey, M.E. and Copeland, J.R.M. (in press). Computerised psychiatric diagnosis in the elderly: AGECAT. *Journal of Microcomputer Applications.*

Department of Health and Social Security (DHSS) (1985) *General Practice Computing, Evaluation of the Micros for GPs Scheme: Final Report.* London: HMSO.

Drury, V.W. (1981) Audit in general practice. *Journal of the Royal College of Physicians, 15,* 259–261.

Evans, A. and Brownbridge, G. (1985) Computer use within the consultation: The Sheffield experience. In M. Sheldon and N. Stoddart (eds) *Trends in General Practice Computing.* London: Royal College of General Practitioners.

Fitter, M.J. (1986*a*) Evaluation of computers in primary health care: The effect on doctor–patient communication. In H.E. Peterson and W. Schneider

(eds) *Human–Computer Communications in Health Care.* North Holland: Elsevier Science Publishers.
Fitter, M.J. (1986*b*) Computers in general medical practice: Educating doctors and receptionists. In S.J. Bostock and R.V. Seifert (eds) *Microcomputers and the Education of Adults.* London: Croom Helm.
Fitter, M.J. and Cruickshank, P.J. (1982) The computer in the consulting room: A psychological framework. *Behaviour and Information Technology, 1,* 81–92.
Fitter, M.J. and Cruickshank, P.J. (1983) Doctors using computers: A case study. In M.E. Sime and M. Coombs (eds) *Designing for Human–Computer Communication.* Academic Press: London.
Fitter, M.J., Evans, A.R. and Garber, J.R. (1985) Computers and audit. *Journal of the Royal College of General Practitioners, 35,* 522–524.
Fitter, M.J., Fox, J., Brownbridge, G., Cruickshank, P.J., Sime, M.E. and Bardhan, K.D. (1983) The doctor's role and the computer in the consulting room. Sheffield: MRC/ESRC Social and Applied Psychology Unit Memo no. 553.
Fitter, M.J., Garber, J.R. and Wall, T.D. (1982) Human factors evaluation of the IBM-SPCS (Report on Phases 1 and 2). Sheffield: MRC/ESRC Social and Applied Psychology Unit.
Fitton, F. and Acheson, H.W. (1979) *The Doctor–Patient Relationship.* London: HMSO
Friedson, E. (1970) *The Profession of Medicine.* New York: Harper and Row.
Hammersley, P. (1975) Principles of control in patient administration. Mimeo, Addenbrooke's Hospital, Cambridge.
Herzmark, G.A. (1986) A study of the impact of computers on general practice administration. Paper presented to The British Psychological Society Annual Conference, Sheffield, available from The British Psychological Society, Leicester.
Herzmark, G.A., Brownbridge, G., Fitter, M. and Evans, A. (1984) Consultation use of a computer by general practitioners. *Journal of the Royal College of General Practitioners, 34,* 649–54.
Hopley, M. (1986) Software review of *Beat the Bottle. Bulletin of The British Psychological Society, 39,* 57.
Knill-Jones, R.P., Dunwoodie, W.M., Crean, G.P. (1985) A computer-assisted diagnostic decision system for dyspepsia. In M. Sheldon, J. Brooke and A. Rector (eds) *Decision Making in General Practice.* London: Macmillan.
Lucas, R.W. (1977) A study of patients' attitudes to computer interrogation. *International Journal of Man–Machine Studies, 9,* 69–86.
Lucas, R.W., Mullin, P.J., Luna, C.B.X. and McInroy, D.C. (1977) Psychiatrists and a computer as interrogators of patients with alcohol-related illnesses: A comparison. *British Journal of Psychiatry, 131,* 160–167.
Marinker, M. (1984) Developments in primary health care. In G. Teeling Smith (ed.) *A New NHS Act for 1996?* London: Office of Health Economics. London.
Nygaard, K. and Sorgaard, P. (1985) The perspective concept in informatics. Mimeo available from the Department of Computing, Aarhus University.
Pringle, M. (1985). Using a computer in the consultation. In M. Sheldon and N. Stoddart (eds) *Trends in General Practice Computing.* London: Royal College of General Practitioners.

Pringle, M., Robins, S. and Brown, G. (1984) Computers in the surgery: The patient's view. *British Medical Journal, 288*, 289–291.
Royal College of General Practitioners (1980) *Computers in primary care: The report of the computer working party*. Occasional paper 13. London: Royal College of General Practitioners.
Shepherd, S. (1984) The Update Primary Care system – good or bad? *Computer Update, 3*, 11–12.
Sims, P. and Herzmark, G. (1986) Attitudinal responses of office staff in General Practice to computer use. Sheffield: MRC/ESRC Social and Applied Psychology Unit Memo no 796.
Skinner, H.A. and Allen, B.A. (1983) Does the computer make a difference? Computerised versus face-to-face versus self-report assessment of alcohol, drug and tobacco use. *Journal of Consulting and Clinical Psychology, 51*, 267–275.
Stott, N. (1985) The place of computers in primary health care. In M. Sheldon and N. Stoddart (eds) *Trends in General Practice Computing*. London: Royal College of General Practitioners.
Tate, P. (1983) Doctors' style. In D. Pendleton and J. Hassler (eds) *Doctor-Patient Communication*. London: Academic Press.

EXPERT SYSTEMS IN THE HEALTH FIELD

Russell Thomas

WHAT IS EXPERT KNOWLEDGE AND WHO IS THE EXPERT?

There exists within the health service a wide range of specialities, from medical consultants to support specialists. Each has its own form of expertise and over the last 20 years there has been an increasing tendency to introduce the computer into these specialities. In such applications the predominant role for the computer has been that of a relatively unintelligent tool in which information is processed using basic algorithms to carry out simple and routine tasks. Nevertheless, such uses have resulted in the general acceptance of computers within health care settings.

More recently a new role for computers has been proposed, in which the problems to be dealt with are no longer mundane. The intention has been to provide a tool which is capable of a much more sophisticated approach to tasks such as complex problem solving and decision making – behaviours that are similar in nature to those of the human expert. Indeed, such a role provides advantages over and above the immediate aid to clinical staff in their decision-making activities. Appropriate systems can function at all levels by providing continuous access to a specialist's expertise without the need for the specialist to be present. It is the development and application of these so-called expert systems and the role that psychology plays in that development that this chapter considers.

Before considering expert systems in detail, a fundamental question must be posed: 'Will it actually be possible to produce such a tool?' If it is possible, how will such systems be used and in what way will it help the health service? As this chapter illustrates, it certainly is possible to produce such tools, and research projects are outlined that range over the skills of health care diagnosis, assessment, support and education.

A further question which also needs to be addressed is to what extent such systems are cost-effective. Unfortunately, if the present systems are used as a guideline, the conclusion would be negative: in general they are not cost-effective. However, it is contended that such a conclusion is premature since the more recent expert systems are much more sophisticated. The system implementers are becoming more skilled at their task, and although the early systems consumed much time and effort, there is now evidence that useful systems can be implemented using comparatively meagre resources. It is largely through understanding the expert's behaviour when making decisions and solving problems that such advances have been possible.

WHAT IS AND WHAT IS NOT AN EXPERT SYSTEM?

Before considering their development and importance within the health care field it is, perhaps, important to begin by defining what is meant by the term 'expert system'. In general almost as many definitions of this term will be found as there are books on the subject. However, the main characteristic of an expert system is that in some way it generates performance similar to that of the expert whose expertise it contains. This is generally done by modelling the expert's inferencing processes.

A more circumspect definition includes the fact that such systems are most useful when at least one of the following characteristics is present in the expert's area of expertise (domain):

- ☐ There is uncertainty about the reliability of information entering the system.
- ☐ There is uncertainty about the relevance of information entering the system.
- ☐ There is uncertainty about the knowledge that exists in the domain.
- ☐ There is uncertainty as to how different information and knowledge about and within the domain is combined in reaching a decision.
- ☐ A set of rules exists to solve a problem but the range of possible solutions is so wide as to make it impractical to test them all.

These characteristics can be summarized as stating that an expert system is best used in situations where the expert uses heuristics (rules of thumb) in coming to a decision.

EXPERT SYSTEMS: THEIR STRUCTURE

Before considering how experts' knowledge and thought processes are transferred to the computer, it is appropriate to understand something of how such systems are created. In essence, an expert system consists of four main components: the user interface, a knowledge base, an inferencing mechanism, and a data base (see Figure 1). Each of these components is further sub-divided into specialist components that deal with such issues as explanations to the enquirer, controlling the system procedure and checking its accuracy and consistency.

There are three main types of expert system architecture, which vary according to the form of decision-making rules used. These are (1) production rule architectures; (2) semantic network architectures; and (3) frame formalisms. Each of these is discussed briefly although, in addition, there are a number of hybrid architectures which combine two or more of these formalisms.

Production rule architecture

These are the most prevalent types of structures in expert systems and use simple *if . . . then* rules of the form:

RULE: IF ≪antecedent condition≫ THEN ≪consequent≫

Figure 2 shows an example of this type of architecture that exhibits the expertise of a consultant on infectious diseases.

IF 1. the infection is primary-bacteremia, and
2. the site of the culture is one of the sterile sites, and
3. the suspected portal of entry of the organism is the gastrointestinal tract,

THEN there is suggestive evidence (0.7) that the identity of the organism is bacterioides.

Figure 2. A prototypical production rule.

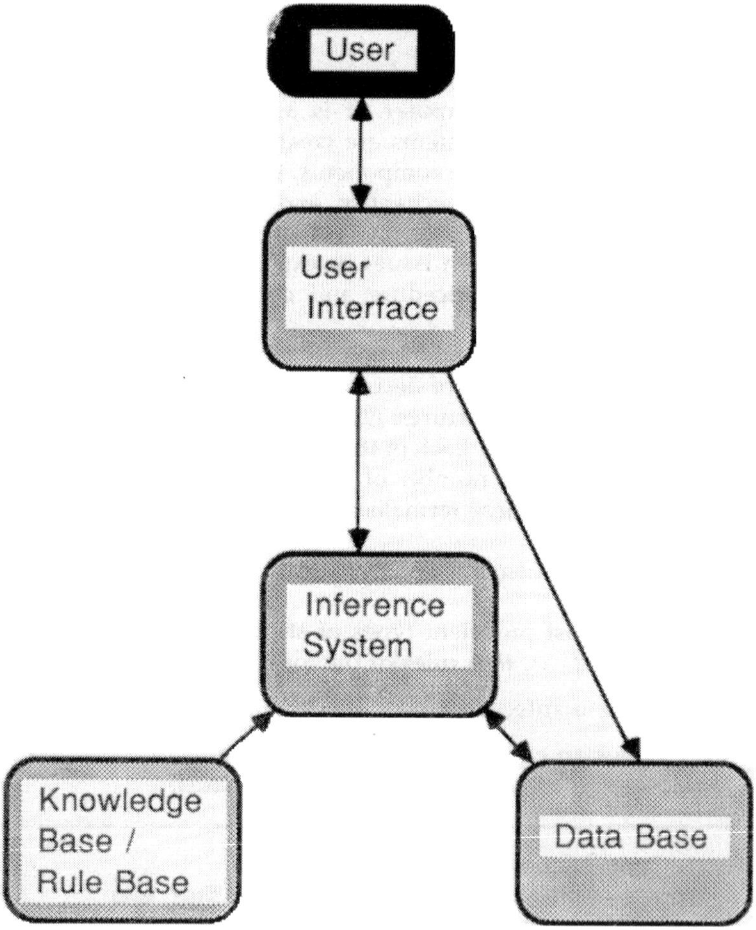

Figure 1. System outline.

As can be seen, the rule in Figure 2 states that if it can be proven that the infection is primary-bacteremia, and the culture comes from a sterile site, and it is suspected that the organism entered via the gastrointestinal tract, then it is likely that the organism can be identified as bacterioides. The use of the statement '. . . evidence (0.7) that . . .' indicates the use of a measure of certainty. In this case it is a statement of likelihood that the conclusion is correct, given the information provided.

There are two standard ways in which the system's interpreter can use the production rules to perform its reasoning task. First, it can chain forward using facts which are provided and applying the antecedent components of the rules to produce a conclusion. The alternative method is to chain backwards from a conclusion (hypothesis) by verifying the facts needed to support it. In many practical expert systems these two interpreter cycles are combined, since there are often control considerations which dictate that some reasoning is best performed by backward chaining and some by forward chaining. However, the predominant reasoning cycle is backward chaining.

Semantic networks

Semantic networks cover a group of systems which share a common notation, consisting of groups of concepts, terms, etc. (nodes) and links between them which indicate the existence of a relationship (arcs). In general, this type of architecture has been developed from work on psychological models of human memory (Quillian, 1968) but, at present, a clear set of unifying principles is lacking for all semantic network systems. Unfortunately, one of the main problems with these systems is that an agreed-upon notion of what a particular representational structure means, is as yet missing. This lack of a formal semantics does not appear in other rule-based structures such as logic.

One of the main terms used in these systems is the *isa* relationship between nodes. For example, to represent the fact that two physicians are human and that Joan is a physician, the structure would be as shown in Figure 3.

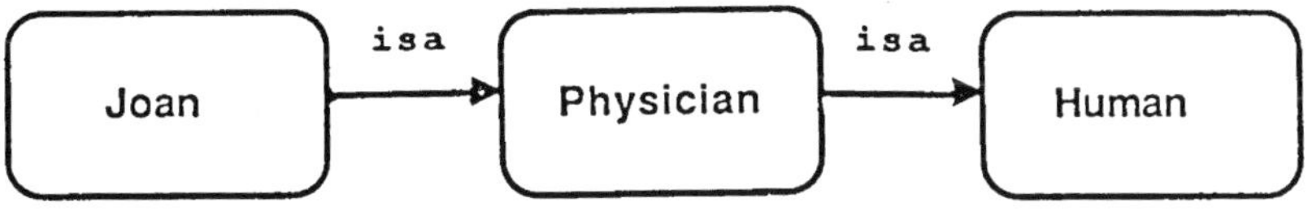

Figure 3. A semantic network.

As these nodes are linked with the *isa* link, the deduction can be made that not only are physicians human, but that Joan is also human. In this way, it is possible to make deductions using this ability to construct inheritance hierarchies.

There are two main ways of using this type of system for making inferences. The first is by using Quillian's concept of spreading activation, in which activity in a set of nodes (possibly constructed as a query) spreads outwards until the activity coincides at a node. At this point the system would attempt to described the connecting route through the net. The second, and most used method, is by matching network structures. In this system a network fragment is constructed to represent the query (see Figure 4), and this is matched against the network data base to see if a match exists.

Figure 4. Structure of the query in a semantic network.

Frames

These are a development of the semantic network structure proposed by Minsky (1975) as a means of understanding complex behaviour. Frames combine features of both of the above architectures, and provide a structure within which new information can be interpreted in terms of existing concepts and knowledge previously acquired.

Reasoning in these systems is achieved by examining the relationship between the generic 'frame' of the problem as a whole and the components from which it is composed (known as 'slots'). For example, an attempt to identify the illness exhibited by a patient would initiate an attempt to fill an 'illness frame'. This, in turn, would initiate attempts to fill the slots (which represent different diseases the system knows about) and each slot would have procedures for establishing its own value (whether the disease exists in the patient). The result of a consultation with such a system is a completed illness frame in which the slots are either false (the patient does not have the disease) or true (the patient has the disease). This is a simplistic description of the reasoning process, and the actual method used will differ from one frame system to another.

MYCIN: AN EXAMPLE OF AN EARLY EXPERT SYSTEM

The prototype of all major expert systems in the field of medicine is the MYCIN system developed as part of the Stanford Heuristics Programming Project (Shortliffe, 1976; Buchanan and Shortliffe, 1984). MYCIN is predominantly a 'consultation' system designed as an aid to decision making in the area of diagnosis and therapy for infectious diseases. Such aids are often required within a hospital system, as the attending physician may not be an expert on the infectious diseases that are apparent or occur during hospitalization. Laboratory analysis of patient specimens (blood, urine, etc) can often positively identify the organisms involved but these analyses often take a considerable time (12 to 48 hours or more). Therefore, the physician must decide the probable causative organism and treatment regime using incomplete information.

As the consultation proceeds with MYCIN, the system builds up a data base of information on a number of contexts, such as the patient who is the subject of the consultation (PERSON), a number of 'current cultures' taken from the patient (CURCULS) and the 'current organisms' which have been isolated from the cultures (CURORGS). The aim of the system is eventually to identify (IDENT) the organism responsible for the illness, with the ultimate goal of recommending a particular drug regimen.

MYCIN is based upon a production rule formalism and uses rules of the form

IF antecedent conditions THEN consequences cf

to contain the knowledge of the domain. A rule may consist of one or more antecedent conditions and one or more consequent conditions. Each of the consequences may also have associated with it a probability-like *cf* (certainty factor) value which represents the degree of confidence that the conclusion is true (a typical MYCIN rule can be seen in Figure 2). The certainty values range between –1 and +1, where +1 implies that the parameter is 'known with certainty' and a value of –1 indicates that the parameter is 'known with certainty not to have that value'. Many of the parameters may have more than one value and each value may be held with a different degree of certainty. For example a CURORGS may have an IDENT which has the values of 'e.coli (.6)', 'streptococcus-group-a (.29)', 'bacteroides (.35)'.

One important feature of the MYCIN system is its ability to provide explanations about its reasoning process and to explain to the user why it requires an item of information. These facilities address two main issues. First, there is the need for the expert to interact with the system and to validate the reasoning steps. Second, the system user (physician) needs the assurance provided by the system's explanation of its reasoning steps before administering a potentially dangerous procedure or drug.

At the conclusion of the consultation, MYCIN provides information on the infections that it has identified and a recommendation of the appropriate drug therapy. Although the MYCIN system is identified as the flagship of medical expert systems, it should always be borne in mind that this system is basically a research tool and is not, in fact, widely used in hospitals.

SYSTEM DEVELOPMENT

Knowledge acquisition

Although there are relatively few psychological theories that can be directly applied to the elicitation of knowledge from domain experts, there are a number of theories and methodologies which are relevant. Gammack and Young (1985) identified three techniques which they claim will aid the elicitation process: protocol analysis, multidimensional scaling, and concept sorting. In addition, there are some clear guidelines derived from investigations of human memory and other areas of cognitive psychology. Finally, the area of social psychology may provide some pointers on how to maintain the expert's interest and motivation to perfect the expert system.

There is a cycle in the development process of an expert system (see Figure 5) which starts with the identification of the main characteristics of the domain. Most developers of expert systems use interviews with the expert at this stage, utilizing either a free or structured interview procedure.

The objective at this stage, is to identify the major domain constructs so that the system developer (Knowledge Engineer) and expert have some common basis of understanding.

Further development and refinement of these initial concepts can be aided by the use of a modified form of Kelly's (1955) repertory grid technique. This technique, part of Kelly's Personal Construct theory, was developed as a clinical tool. However, recent research has shown it to have a role in the primary development of expert systems (Boose, 1984).

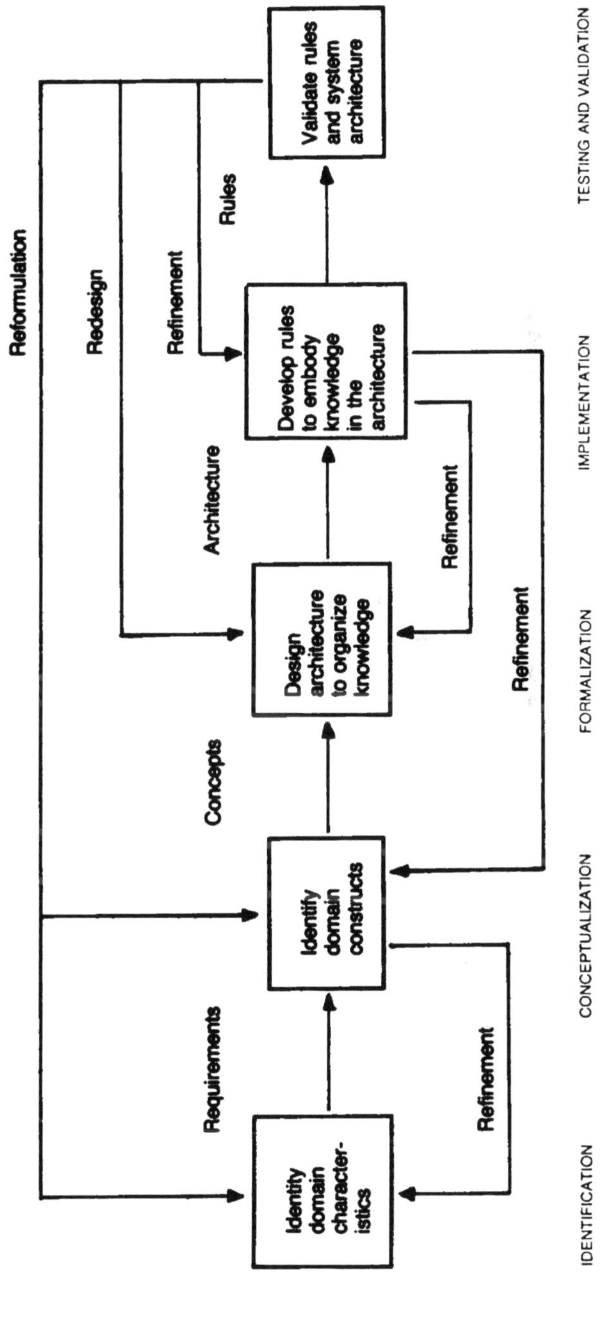

Figure 5. Stages of system development

Boose's (1984) Expertise Transfer System is a computer program which conducts a primary interview with a domain expert and then constructs and anaylyses interactively a construct grid. The grid analysis provides sufficient information about the elements and constructs in the domain to enable a set of prototype rules to be generated by the system. While this approach provides for rapid prototyping of a simple expert system, there is a general need for rule refinement and the addition of control structures.

The knowledge gained by the use of the grid technique is predominantly declarative knowledge and does not provide information of a heuristic or procedural nature. TEIRESIAS (Davis, 1982) is a system which aids the refinement and development of an expert system and as such would provide a means of gaining access to the missing forms of knowledge. Other, non-automated techniques for gaining this type of knowledge include protocol analysis, in which the process of problem solving by an expert is monitored and the expert's solution methods are analysed to develop a set of production rules which represent their behaviour.

When attempting to elicit information from the expert, it is important to remember that providing an appropriate context for recall will facilitate the expert's access to the information. The development of an early prototype system can provide the expert with a context in which to further structure knowledge and to identify the knowledge missing from the system. For when a program performs wrongly, it often becomes plain to experts that they have overlooked some important information or failed to consider the limitations on the context of some specific item of knowledge.

A number of psychologists have recommended spending an initial period with the experts helping them to structure their own knowledge prior to attempting any formal knowledge elicitation. It is important to point out to the experts that they need to be able to justify their reasoning steps and to identify an appropriate level of detail for the system to operate at. Experts, aided in this way, do seem to provide a more complete prototype system than those experts who do not go through such a prior structuring exercise.

Most expert systems have provided a model of expertise at a relatively shallow level, based upon the expert's verbal reports. This is prone to error and in many tasks it is necessary to reason from first principles (Davis, 1983), whereas the expert will often avoid such low level descriptions and represent relationships at a more abstract level. The need for access to deep knowledge (first principles) sources is becoming evident (cf. Clancey, 1983), particularly where the knowledge may be used subsequently in tutoring systems.

A further area of difficulty is that of uncertain knowledge and reasoning using such knowledge. Some systems have adopted Baysian statistics for representing uncertain knowledge of this type whereas others have adopted MYCIN type 'confidence factors'. In the case of Baysian statistics it is necessary for the expert to provide accurate estimators of the prior probabilities of events and if the data are not available then the recourse is often to rely upon the expert's judgement. Unfortunately there is considerable psychological evidence (Tversky and Kahneman, 1977) to indicate that human judgement of probabilities is prone to considerable error. The use of the less formal 'confidence factors' does not get completely away from this problem, but Fox (1983) has suggested that the use of category scales can produce more precise judgements of certainty from experts.

System design

The need to match the expert's own representation of the domain knowledge to the architectural formalism of the domain is evident. The predominance among early expert systems for a production rule formalism is in accord with the observation that experts find it relatively easy to express their knowledge in this format.

Later systems move away from such a uniform structure towards a more catholic approach, in which a single system may utilize two or three different formalisms (cf. Aikins, 1983). To some extent, this is a result of the increasing need for systems to have access to the deeper (first principles) knowledge structures of the expert. Such forms of knowledge will be needed by the system only if other higher level rules fail to be applicable and, therefore, a concept of levels of knowledge sources needs to be encoded in the architecture. In a similar way, some architectural features provide more efficient and accessible control structures for the knowledge sources.

When implementing an expert system, the need to select an appropriate architecture is important and ignoring this stage probably accounts for a large number of expert systems which have never progressed beyond the prototype stage. This tendency is exacerbated by the advent of relatively inexpensive and limited expert system shells which provide only a restricted and constrained architecture. Such systems are useful for prototyping and for use in limited domains but in cases where the problem is more complex than a simple tree structure, they are generally inappropriate. Sophisticated expert shells with multiple architectures are currently available but are often expensive (over £100,000), need specialized hardware to run on (often over £50,000) and demand experienced personnel. The alternative is to build a system from scratch. This approach has a longer development time, with the associated costs, but allows greater flexibility and enables the system's structure to reflect special needs.

Evaluation and user acceptance

An expert system is worth developing only if it is able to perform at an acceptable level. The problem is evaluation: how do we evaluate the system's performance? What do we evaluate it against? The usual answer to these questions is that the system's performance on a set of cases is compared with that of a panel of human experts. Again though, what do we compare? Is it the final conclusion that both the system and expert panel agree upon, or is it that the reasoning steps of both must agree? What if the panel fails to agree and if the system agrees with the answers of half the panel?

There are no obvious answers to these questions, but they reflect some of the outcomes and criterion problems that have occurred. Problems also occur with the use of a preselected set of test cases. The system may become tuned to the set of test cases and perform at or above expert level, but when given another set of cases, fails to achieve such high levels of performance.

The existence of an expert system does not ensure its acceptance by users. Considerable effort has been made to provide explanation facilities within these systems, for users are unlikely to accept certain types of conclusion without explanation of the logical development of that conclusion. When explanations are provided, they should be at an appropriate level of detail for the user. An expert in a domain would not require the same type of detailed explanation as a non-expert user.

EXPERT SYSTEMS IN MEDICINE

The following is a selection of expert systems which may be found currently in the area of medicine. Many of these have not progressed beyond being a research tool, but it is important to understand the broad spectrum of uses that this technology is capable of addressing.

You will note that none of the systems described has been developed in the UK. In my opinion, this is not due to a lack of interest in this country, but rather to a desire to evaluate the potential of the systems before attempting full implementations. This is best exemplified by the pioneering work carried out by the Imperial Cancer Research Fund (Alvey, Myers and Greaves, 1985; Fox, 1983) in their development of a system to aid in the identification of leukaemia from laboratory tests. There are now though a number of medically-oriented expert systems under development in the UK but they are generally at an earlier stage of development.

Systems in diagnosis, assessment and support

CASNET (Causal ASsociational NETwork). This system, developed at Rutgers University, operates in the domain of diagnosing glaucoma and aids the clinician in the development of treatment therapies. Information within the system is represented as a set of causal associations using a semantic network formalism. The system's knowledge base contains information about the relations between patient's symptoms, test results, disease states and treatment plans. Within the system, there are three planes of knowledge: disease categories, pathophysiological states, and observations. The plane of observations contains nodes which represent the information gathered from the patient (signs, symptoms and laboratory tests) and from these nodes extend links to the pathophysiological plane. In this way pathophysiological nodes are supported by the evidence provided at the observational plane. They, in turn, provide supportive evidence for disease states at the level of the disease category plane. Each of the links between observations and states has associated confidence values which reflect the degree to which the evidence supports the associated states. At the level of disease categories, a disease is defined as a set of confirmed and denied pathophysiological states which are linked via the classification links from the pathophysiological states. When all observations have been entered, the classification tables at the disease categories level are used to determine diagnosis and treatment. The program's performance has been evaluated by opthalmologists and is considered to operate close to expert level (Weiss, Kulikowski and Safir, 1977; Weiss *et al.*, 1978).

PUFF. This system diagnoses the presence and severity of obstructive airway disease (OAD) in a patient by interpreting the measurements from respiratory tests administered in a pulmonary function laboratory. PUFF (Aikins, Kunz and Shortliffe, 1983) uses its knowledge about the types of test results produced by the various pulmonary disorders in diagnosing the patient's disease state. The data used by the system include the patient's history, referral diagnosis and the results of a battery of pulmonary tests (residual volume, total lung volume, changes in resistance, etc). PUFF was developed using the EMYCIN shell (van Melle, Shortliffe and Buchanan, 1981) at Stanford University and tested at the Pacific Medical Center in San Francisco, where its performance was considered satisfactory. Further development of PUFF was carried out by Aikins (1983). This resulted in the CENTAUR system which is seen to address some of the knowledge representation issues which occurred when modifying the knowledge base of PUFF.

CADUCEOUS. This system assists physicians in diagnosing diseases in the area of internal medicine. Its precursor was the INTERNIST-I system (Miller, Pople and Myers, 1982) which attempted to model the clinician's diagnostic reasoning process, particularly in situations where a patient may have multiple disease states simultaneously. The CADUCEOUS program (Pople, 1984) determines a diagnosis through inferring a set of hypotheses which are adequate to explain the patient's state. Each hypothesis is a disease and the system functions by establishing a disease, or set of diseases, which best fits the patient's data. Patient data, in this case, consist of the patient's history, symptoms and the results of laboratory tests. This is one of the largest medical expert systems developed to date and contains profiles of over 500 diseases described by over 3,500 manifestations of disease. Knowledge representation in this system consists of a disease tree which uses the "form-of" relationship to link the various nodes together. Each branching, from a node, represents a refinement of the disease, while terminal nodes are individual diseases. Each disease is associated with a set of relevant manifestations. When a consultation begins, a list of the patient's manifestations is entered and these become associated with the relevant disease nodes. In this manner at the end of the initial data entry phase, the system attempts to refine the diagnosis by partitioning the disease tree into disease areas which together account for all the manifestations. These are then further refined and partitioned until the manifestations are accounted for by a set of terminal nodes (diseases). The system's performance has been described as being equal to that of an expert in complex cases involving multiple diseases.

Drug Interaction Critic. Developed at the Virginia Polytechnic Institute State University (Roach *et al.*, 1984), the critic helps physicians plan the administration of new drugs to patients who are already on a drug regimen. The system's knowledge is represented as a frame structure which includes information on specific drugs' actions, their affinities, storage sites and interaction characteristics with regard to other drugs. Information on the interaction characteristics is organized by frames according to one of four types of mechanism: chemicophysical, pharmacodynamic, pharmacokinetic and physiologic. Using this knowledge and patient data, the system identifies both adverse and beneficial interactions, explains why the interactions occur, answers questions about the interactions, and suggests corrective action for adverse effects.

PEC (Primary Eye Care system). This simple system was developed at Rutgers University (Kastner *et al.*, 1984) as a support aid for primary health workers who need to diagnose and treat common and potentially blinding eye disorders. The system's knowledge, in the form of forward chaining production rules, is based upon the World Health Organisation's guide to primary eye care. There is a simple user interface which predominantly uses yes/no questions in gathering patient data. This system has been coded in BASIC to run on microcomputers.

VM (Ventilator Manager). VM (Fagan *et al.*, 1984) monitors the postsurgical progress of cardiac patients in an intensive care unit (ICU), who require mechanical breathing assistance. The VM system, based upon a production rule architecture, uses both information about the patient (sex, age, etc.) and information from the ICU patient-monitoring system which it interrogates at 2 or 10-minute intervals. Information gained from the ICU monitor includes heart rate, blood pressure and oxygenation. From these inputs, the system identifies possible alarm conditions and changes in the patient's status which require a change in the ventilator therapy. An important aspect of the VM system is its ability to deal with time-varying data and this is partially achieved by organizing the rules into several types: initialization rules, status rules, therapy rules and transition rules. On each occasion that the VM receives an update from the ICU monitor, it reruns all of its rules. The transition rules, for example, are used to decide when a patient's status has changed from the previous occasion that the rules were run. VM's performance has been monitored for some time by research clinicians at Stanford University Medical Center.

Data collection and record management

PIP (Present Illness Program). PIP interweaves the process of patient history taking with that of diagnosis (Kulikowski, 1980). Its objective is to assist a physician to assess the present illness of a patient with oedema. Using a frame type knowledge structure, PIP contains knowledge about prototypical findings such as laboratory data, signs, symptoms, temporal aspects of an illness, and rules for matching a patient's profile against hypothesized diseases or states. Dialogue control is achieved by using a set of diagnostic hypotheses suggested by the patient's complaints to interweave the history taking with the specific information needed to develop a diagnosis. A rank ordering of the disease hypotheses is achieved by the use of certainty measures based upon the ability of the hypotheses to account for findings and the degree of fit between them.

PATREC (PATient RECords system). This system addresses the physician's need for sophisticated access to data bases of patient records (Mittal and Chandrasekaran, 1980). It was developed in the context of the MDX expert system (Chandrasekaran and Mittal, 1983), for the diagnosis of the liver syndrome, cholestasis. In the PATREC system, the user can enter patient data records, which are stored. The user can then access these data using a query language and the system will offer suggestions helpful in diagnosis and prepare summary reports. Knowledge, which is represented within the system as frames, is of two kinds. First, a patient model consisting of patient histories and clinical episodes, and second, a conceptual model of medical data, such as the expected values and significance of laboratory tests. The system is able to perform all of the necessary bookkeeping functions and to perform sophisticated automatic temporal inferencing.

Education

GUIDON. This is a rule-based system for teaching diagnostic problem-solving skills to medical students (Clancey, 1983). It is based upon the infectious-disease diagnostic rules contained in the MYCIN expert system although it is totally separate from the MYCIN system. In teaching, the system selects a case and solves it using the MYCIN system. Then it presents the case to the student to solve and analyses the student's responses and queries during the solution process. Discrepancies between the student's demonstrated knowledge and that used by the system are used to guide the system's tutoring and explanation facilities. One of the problems found with the use of the MYCIN rules was that they provided explanation for a decision at one level and students often required a more detailed explanation including the fundamental principles underlying the application of a rule. Even so, GUIDON's ability to provide a flexible mixed-initiative dialogue with the student resulted in a useful system for tutoring in antimicrobial therapy.

HEADMED. Developed as a tutorial and consulting aid, HEADMED provides support in the area of drug prescribing for psychiatric disorders (Heiser, Brooks and Ballard, 1978). The system uses, in its inference process, knowledge about differential diagnoses of the major affective disorders, organic brain disorders and schizophrenia. In addition, the knowledge base contains rules dealing with behaviour disorders, neuroses, the Minnesota Multiphasic Personality Inventory (MMPI) scales and substance abuse. Using this knowledge, the system is able to advise physicians on the diagnosis of a broad range of psychiatric disorders and to provide treatment recommendations. These recommendations are based on the psychopharmacology requirements (for example drug dosage and duration of therapy) of the patients. The system was developed at the University of California, Irvine, using the EMYCIN shell.

THE WAY FORWARD

It is evident that there is a role for expert systems in the health service. At present, the research has tended to focus on areas where there are high levels of expertise. This is true not only in medicine but in all other areas where expert systems have been applied. There are still a number of untapped areas within the health field. Psychological testing is one example where the application of expert system technology would have a high pay-off potential.

In the future, a number of expert, or knowledge based, systems are likely to become incorporated within other systems. For example, EEG, ECG and intensive care patient monitoring systems will be available with integral analysis and interpretation facilities based upon these types of technologies.

Even so, without the developing understanding of the structure of human knowledge and expertise provided by psychology, there is a likelihood that this whole endeavour to capture human expertise within a computer system would founder. It is also clear that there is considerable scope for further psychological research within this field including the development of knowledge elicitation techniques.

REFERENCES

Aikins, J.S. (1983) Prototypical knowledge for expert systems. *Artificial Intelligence, 20*, 163–210.

Aikins, J.S., Kunz, J.C. and Shortliffe, E.H. (1983) PUFF: An expert system for interpretation of pulmonary function data. *Computers and Biomedical Research, 16*, 199–208.

Alvey, P.L., Myers, C.D. and Greaves, M.F. (1985) An analysis of the problems of augmenting a small expert system. In M.A. Bramer (ed.) *Research and Development in Expert Systems*. Cambridge: Cambridge University Press.

Boose, J.H. (1984) Personal Construct theory and the transfer of human expertise. In T. O'Shea (ed.) *ECAI-84: Advances in Artificial Intelligence*. Amsterdam: Elsevier Science Publishers.

Buchanan, B.G. and Shortliffe, E.H. (eds) (1984) *Rule-Based Expert Systems: The MYCIN Experiments of the Stanford Heuristic Programming Project*. Reading, Mass.: Addison-Wesley.

Chandrasekaran, B. and Mittal, S. (1983) Deep versus compiled knowledge approaches to diagnostic problem-solving. *International Journal of Man–Machine Studies, 19*, 425–436.

Clancey, W.J. (1983) GUIDON. *Journal of Computer-Based Instruction, 10*, 8–15.

Davis, R. (1982) TEIRESIAS: Applications of meta-level knowledge. In R. Davis and D.B. Lenat (eds) *Knowledge-Based Systems in Artificial Intelligence*. New York: McGraw-Hill.

Davis, R. (1983) Reasoning from first principles in electronic troubleshooting. *International Journal of Man–Machine Studies, 19*, 403–423.

Fagan, L.M., Kunz, J.C., Feigenbaum, E.A. and Osborn, J.J. (1984) Extensions to the rule-based formalism for a monitoring task. In B. Buchanan and E. Shortliffe (eds) *Rule-Based Expert Systems: The MYCIN Experiments of the Stanford Heuristic Programming Project*. Reading, Mass.: Addison-Wesley.

Fox, J. (1983) *Formal and Knowledge-Based Methods in Decision Technology*. London: Imperial Cancer Research Fund Laboratories.

Gammack, J.G. and Young, R.M. (1985) Psychological techniques for eliciting expert knowledge. In M.A. Bramer (ed.) *Research and Development in Expert Systems*. Cambridge: Cambridge University Press.

Heiser, J.F., Brooks, R.E. and Ballard, J.P. (1978). Progress report: A computerized psychopharmacology advisor. Proceedings of the Eleventh Colloquium International Neuro-Psychopharmacologicum, Vienna, Austria.

Kelly, G.A. (1955) *The Psychology of Personal Constructs*. New York: Norton.

Kastner, J.K., Chandler, R.D., Weiss, S.M., Kern, K. and Kulikowski, C.A. (1984) An expert consultation system for frontline health workers in primary eye care. *Journal of Medical Systems, 8*.

Kulikowski, C.A. (1980) Artificial intelligence methods and systems for medical consultation. *IEEE Transactions on Pattern Analysis and Machine Intelligence, PAMI-2(5)*, 464–476.

Minsky, M. (1975) A framework for representing knowledge. In P.H. Winston (ed.) *The Psychology of Computer Vision*. New York: McGraw-Hill.

Miller, R.A., Pople, H.E. Jr and Myers, J.D. (1982) INTERNIST-I, an experimental computer-based diagnostic consultant for general internal medicine. *New England Journal of Medicine, 307*, 468–476.

Mittal, S. and Chandrasekaran, B. (1980) Conceptual representation of patient databases. *Journal of Medical Systems, 4*, 169–185.

Pople, H.E. Jr (1984) Knowledge-based expert systems: the buy or build decision. In W. Reitman (ed.) *Artificial Intelligence Applications for Business*. Norwood, NJ: Ablex.

Quillian, M.S. (1968) Semantic memory. In M. Minsky (ed.) *Semantic Information Processing*. Cambridge, Mass.: MIT Press.

Roach, J., Lee, S., Wilcke, J. and Ehrich, M. (1984) An expert system that criticizes decisions in combination drug therapy. Proceedings of the First Conference on Artificial Intelligence Applications. IEEE Computer Society.

Shortliffe, E.H. (1976) *Computer-Based Medical Consultations: MYCIN*. New York: American Elsevier.

Tversky, A. and Kahneman, D. (1977) Judgement under uncertainty: Heuristics and biases. In: P.N. Johnson-Laird and P.C. Wason (eds) *Thinking: Readings in Cognitive Science*. Cambridge: Cambridge University Press.

van Melle, W., Shortliffe, E.H. and Buchanan, B.G. (1981) EMYCIN: A domain-independent system that aids in constructing knowledge-based consultation programs. *Machine Intelligence, Infotech State of the Art Report 9, 3*, 249–263.

Weiss, S.M., Kulikowski, C.A., Amarel, S. and Safir, A. (1978) A model-based method for computer-aided medical decision-making. *Artificial Intelligence, 11*, 145–172.

Weiss, S.M., Kulikowski, C.A. and Safir, A. (1977) A model-based consultation system for the long-term management of glaucoma. *Fifth International Joint Conference on Artificial Intelligence*. Los Angeles, Ca.: Morgan Kaufmann.

THE DISABLED

Gerald W. Hales

The effects of a disability on individuals either prevent them from doing certain things or make those things very difficult to do. These 'certain things' fall into a number of categories, but most people recognize the obviously practical: walking, talking, hearing, seeing. There are, however, some equally practical but less obvious functions that will be of great importance to the individual concerned: into this category fall such things as heart function, urinary function, nerve transmission, blood cell construction. Difficulties in such functions may make it difficult for the individual to carry out certain tasks or they may make it almost impossible to do very much at all. The sufferer from angina or haemophilia, for example, must take care in many aspects of life.

The applications of many of the new developments in information technology relate in a direct way to a physical function (such as opening the curtains, dialling on a telephone), but they also relate in a less direct, although very important and powerful way to the ability of the individual to be independent, make personal decisions and carry them out. Technological developments will contribute greatly to the psychological well-being of the disabled person, and are not to be considered only as mechanized full-time body servants.

Hawkridge, Vincent and Hales (1983) state:

> They [disabled people] are often left isolated, powerless and dependent. They are deprived of important ways of expressing their individuality (p.3).

This aspect of disability is affected by a third set of functions which may be rendered difficult for the disabled individual and not infrequently these represent a very great handicap indeed in terms of any interface with the non-disabled world. These are the functions of learning and communication. A difficulty in either of these spheres (and they are not entirely independent) produces a considerable obstacle for individuals in the development of a personal framework for life, and often renders them powerless to determine their own pathway through life, leaving them isolated and dependent on others. People with a communication handicap will have great difficulty in developing and projecting their own personality; and a problem in receptive communication makes learning difficult, which in its turn deprives them of the ability to gain access to the basic units of empathy and unity with their fellow human beings.

So the new information technology may be seen as a potential for bridging gaps in two ways: one enables the individual to control a mechanical device which substitutes a physical function (such as steering a wheelchair), while the other enables the individual to gain access to a system which would not otherwise be accessible (such as 'reading' a book that cannot be seen).

Either of these may improve the individual's potential for communication or learning, but the second approach is more likely to do so. Thus a boy provided with a sophisticated radio hearing aid has the practical function of hearing improved; but, significantly, his mother is reported in a newspaper as discussing the benefit in terms of learning potential and communication: 'It makes such a big difference to John. He can *understand* more of what is going on and it also *improves his speech*' (Beds and Bucks Advertiser, 1983).

Thus it is clear that the factors to be considered when using the new technology with disabled people include the human factors to a very large degree. This is not quite the same with the non-handicapped, as computers have begun to develop in ways which increasingly minimize the problems.

Duncan (1983) writes:

> Some [personal computers] are designed to be friendly, even helpful. If we have good eyesight, good fingers, and a fairly functional brain, we can hardly ask for more . . . The conventional computer terminal, or the personal computer with VDU and printer, makes, however, no concessions to the operator who through physical disability cannot use a standard typewriter keyboard or read the characters on a screen (p.2).

There are, then, two sets of problems:

1. How can technology assist disabled people do what they want?
2. How can the system be adapted so that disabled people can use it in the first place?

This chapter examines how disabled people are assisted in their attempts to achieve their own goals (whatever they might be) by the new information technology, and looks at the following areas of design and application: matching to the individual; human functioning; communication; education; language; personal expression; information; employment.

MATCHING THE EQUIPMENT TO THE NEEDS OF THE INDIVIDUAL

The system adaptation can relate either to the need to modify the hardware, for example where a physically disabled person simply cannot push the keys, or to the need to adapt the person–machine interface, which is the case with a blind operator who can use the keyboard, but cannot see the screen.

The problem may be no more complex than that of those who cannot control their fine motor movement sufficiently to prevent themselves pressing two keys at once, or it may involve someone who cannot use a keyboard at all. Often the question of matching the equipment to the needs of the individual is very complex. The Trace Research and Development Center for the Severely Communicatively Handicapped in Wisconsin (USA) faced this problem in 1983, and the outcome was a detailed study which compared microcomputers in their applicability to people with handicaps. They developed a checklist system for analysing needs, which is very useful in making the primary decision of which microcomputer to choose (Vanderheiden, 1983).

The items considered by the Trace Center include such things as:

- Are the keyboard, and the keys, the right size?
- Can the keyboard be detached from the bulky parts?
- Can other devices be easily connected?
- Is the screen layout suitable and is the display appropriate?
- Is it easily portable?
- Is the memory large enough?
- Is suitable software available?
- What are maintenance facilities like?

There are many more factors to be considered, and the Trace team states that it is important to define the need clearly as a first step. Duncan (1983) makes this clear in his example of a man whose speech was greatly impaired, together with loss of use of his right side, following a stroke. This man can still type, but only with his left hand, so the technology was modified for one-hand use, so that by typing in a brief code, the machine speaks for him. He does the coding himself, and has no difficulty programming or using the machine. There was a physical problem of hitting more than one key at a time, however, and the device had to be fitted with a key guard.

Many disabled people who need help in carrying out the physical parts of life are very weak as a result of the condition which causes the disability. This means that they have difficulty in operating some mechanical devices which could offer them help because they are not strong enough to do so. Here is an area in which the new technological advances can be of great help, for the computer may be operated with very little strength (even, as is mentioned later, by something as minimal as muscle flexion and the resultant voltage changes). In its turn, the computer may control equipment capable of carrying out large-scale movement.

One device which has been very successful in this area is the Microwriter, a small keyboard substitute which can be operated with one hand and requires only extremely light pressure to operate it. Originally conceived as a note-taking device, it can now be used as a computer input terminal, and has been modified for many different types of user, including the left-handed.

The principles of devices such as the Microwriter make it possible for a wide range of people to have access to the control functions of computers, including some groups which not only cannot use mechanical equipment, but often cannot use conventional computers either. This includes such people as those suffering from haemophilia, where the pressure of pushing keys can cause bleeding, those whose ability to move their arms is restricted (which affects keyboard use as much as difficulties with hands) and individuals who must use unusual parts of their bodies to push any sort of key.

DESIGNING FOR HUMAN FUNCTIONING

There is, of course, a profound difference between the part a microcomputer plays in the life of someone able-bodied, and the part it plays in the life of a handicapped person. For the non-handicapped person, it is a device which may be very helpful, but which can be used whenever needed and left alone for the rest of the time. For the disabled person, however, it is not an *addition* to life, but actually *part* of it. Disabled people rely on it to carry out some of the functions that everyone else is able to do without the technology. This means that factors such as size, portability and conspicuousness are very important.

The non-disabled person may well perceive the handicapped person as 'different' in a way that implies being 'less' – there is something that the handicapped individual cannot do which everyone else apparently can. This is often true in the literal sense that the individual may have some function or part of the body actually missing. However, there are two aspects to the ability to 'do things'. One is the function of the body, mind and brain which in combination enables us to participate in activities, while the other comprises the physical action involved in carrying out the process. These we may call the *human function* and the *physical function.*

Thus the ability to hear is a human function which enables us to carry out physical functions like listening to a symphony. The ability to speak is a human function which enables us to carry out tasks such as buying vegetables (a physical function).

The self-image of the disabled relates to how they *are,* not how they were or how they might be. This is because their frame of reference does not include those possibilities of human function which are, in reality, impossible: you do not incorporate walking into the functional sub-set of your self-image if you have no legs, nor seeing if you have no vision.

However, this does not necessarily apply to the physical function related to the human function: paraplegic people may comprehend completely that they will never walk, but still have a need for mobility. Blind individuals may understand perfectly well that they cannot see, and therefore exclude vision from their frame of reference, but still wish to be able to read. Thus the technology replaces the human function in the individual, enabling the physical function to be carried out.

It will be seen easily how this involves a need for special considerations in the construction of equipment, for that which is large, bulky, difficult to use or conspicuous acts as a signal to handicapped people and others that constantly labels them as 'different', often with the implication that this means being somehow 'less'.

The difficulties which can arise from such a situation are made quite clear by a story from someone who cannot speak and uses a synthetic voice producer. He reports a visit to a restaurant for a meal which was considerably disrupted by the waiter once he discovered that his customer used an artificial voice. The handicapped person,

Michael Williams, records his despair that all the waiter wanted to do was 'play a little game of "Can you say this on that machine?,"' (Williams, 1981). Williams makes it clear that this is not an unusual or strange occurrence once people discover for the first time how he normally communicates.

From this incident, we can see clearly the differences in perception between the two participants in the act. To the waiter, the voice synthesizer was a 'machine' which could do something interesting and was fun. To Mr Williams, however, the device was his voice – as much part of him as the voice system used by everyone else. Thus, to treat it on the level of a game was a considerable invasion of his concept of himself.

In many ways the technology directly affects the physical functioning of the individual, for this is the part that is handicapped. In very many cases, though (possibly even all cases), there is a profound effect on the psychological well-being of the individual concerned. The same Michael Williams just mentioned actually gave a speech (and remember, he has no voice) to the Rehabilitation Engineering Conference at Stanford University in 1982, during which he made the following remarkably descriptive statement of what effect the technology had on his state as an individual:

> If, when I was a little boy, someone had told me that I would grow up and make speeches to large groups, I would have called him either a fool or a madman. Yet here I am. I can only say this: modern technology has allowed me to release my creative spirit where it can soar, free, high above the clouds. Without the fruits of modern technology I would probably be stuck in a room counting the hours until my death. To some people, this synthesiser may be an ugly box with cables. To me, however, it is an analogue for freedom. Let freedom ring (Williams, 1982).

People vary a great deal, of course, in their ability to incorporate such pieces of technology into their life. Duncan (1983), reporting work on synthetic voices at Bristol University, writes of the very great differences in the ways people react to what the technology will do:

> One is happy to take his machine to the shop around the corner; he doesn't care what it sounds like as long as he is understood. Another will come to terms, reluctantly, with the machine only when it produces a voice as clear and euphonious and feminine as the voice she has lost and when it is small enough to be concealed (p.5).

DESIGNING FOR COMMUNICATION

Devices such as artificial voices, which relate to the communication function, are by definition at their highest level of usage in interactive and, therefore, to some degree public situations. Indeed some pieces of equipment are only of value in such a context. However, it is also easy for the handicapped user to believe that the technology is more conspicuous than it really is, at least in the perceptions of others.

Hales (1978) reports an investigation of the use of a new hearing aid specifically designed for partially-hearing students in a tutorial situation. Although the major thrust of the experiment was to consider the effect of the equipment on communication, those taking part were asked for their reactions to the practical aspects of usage: 62.5 per cent of the hearing-impaired students thought the device was 'conspicuous in use', compared with only 23.8 per cent of the normally-hearing students.

A major need for anyone suffering from a handicap, therefore, is that of *communication*. This is most obviously true for those suffering from certain sensory and communicative disorders – the blind, the deaf, those with speech impairments. However, it is just as true, although possibly in a different way, for the physically handicapped. Many physical handicaps make communication difficult: some directly affect relevant abilities, such as speaking, hearing or seeing, and this is frequently the case in progressive diseases. The difficulties of such a situation can be compounded by the physical components of the disability, as those with some conditions are very weak, and may find it either impossible to operate the equipment which could otherwise assist them to communicate, or become exhausted at a very rapid rate.

Even in less complex and less handicapping conditions communication is affected: paraplegic people who are still able to speak, hear, see, use hands, etc., despite loss of use of their legs are still prevented from exercising the normal functions of personal contact, going where they will, meeting whom they will and communicating however they wish. Thus such individuals may well find that a good deal of communication takes place, of necessity, through such devices as the telephone, with a concomitant loss of or reduction in the quantity and quality of the communication process.

John Eulenberg is professor of linguistics, African languages and computer science at Michigan State University (USA), and has developed a view that people have not just a need, but a right to communicate, for without this ability they are excluded from much that is human. It was for this reason that he developed Voice Output Communication Aids (VOCAs) at the Michigan Artificial Language Laboratory, and his work has covered a wide range of problems.

One device was developed at Michigan for Jim Brooks, who had spastic athetoid cerebral palsy at birth. He could not communicate, and everyone except his parents assumed he was retarded. His parents realized he was not, because he learned to communicate by the only means he could handle – typing on an electric typewriter with his big toe. After the development at Michigan of a speaking computer system which is mounted on his wheelchair, not only can Jim communicate with people, but it has been discovered that he has a brilliant scientific mind (Moses, 1983).

Jim is still able to use only his right foot, as it is the only part of his body he can control. He uses a joystick to operate the computer, calling up words or parts of words from the memory: these are formed into sentences by Jim, who can see what he is doing on an eye-level screen, and once the sentence is complete he uses a synthetic voice to speak it.

One innovation which has been the direct result of the new technology is the ability to use parts of the body which are not normally part of the communication process to facilitate communication. This situation arises because many physically handicapped people have very limited amounts of control over their bodies, and these areas do not always include the normally 'useful' parts, such as arms, hands and feet. They will find it difficult, therefore, to use any equipment at all, whether meant for communication or otherwise.

It is now increasingly possible to design interfaces so that operation of equipment is possible through those muscular areas that can be controlled adequately. In this way, it was reported on a television documentary that Dick is experimenting with electrodes attached to muscles (BBC, 1984); the electrodes pick up the voltage levels in his muscle cells and thus enable control to be established. With electrodes attached to several muscles, it would be possible to send quite complex coded signals to the computer. Dick is also advised by Professor Eulenberg.

DESIGNING FOR EDUCATION

All the situations in which disabled people are able to compensate for their disability (to a greater or lesser extent) by using complex computer-based equipment have one underlying component in common:

they assume that the individual is capable of *understanding* what to do. This is also true of other areas of endeavour for disabled people, such as trying to gain and keep employment. A major factor in this is the access of the person to education, not only in the limited sense of formal education at school, but in every aspect of learning, from birth.

A disability which is present at birth or occurs during childhood is of course highly likely to affect the individual's education. This is particularly true of those people with disabilities which include some communication component. However, the final effect of this situation is very serious, for the disabled child not only grows up into a disabled adult (with a few exceptions) and has to face the rigours of adult life with the extra task of coping with the problem, but also has to face them with the additional burden of an inadequate education. Thus the individual becomes doubly handicapped – once by the disability and once by the education deficit arising from that disability.

In the field of the education of disabled people, the new technology can play a substantial part. Although a good deal of what is done relates very closely to contexts outside education, such as carrying out tasks, establishing mobility and enabling communication to take place, much of the new technology – both hardware and software – has been written, adapted or used with education specifically in mind.

Disabled people may well have problems with some of the basic facets of educational life. They may be able to write only very slowly or with great difficulty or even be unable to do so at all. They may be unable to see written material, as in the case of a visually-impaired person, or unable to perceive it properly, as in the case of some language disabilities; or they may be unable to communicate with teachers and other students because of a difficulty in hearing or speaking.

In all of these cases, both good teachers and specialist technology can help. There is, though, a deeper aspect to their learning experience, for students are often unable to participate in some of the reinforcing work which is a substantial part of practising skills in education. Foulds (1982) has offered the suggestion that many are debarred from much of the practice of spelling, or mathematics, because of an inability to do 'rough work', or work things out visually on paper. In the same way, taking notes is impossible for many.

These difficulties do not affect only the physically disabled. In California, Project Video Language has worked on ways to make repetitive classroom drill more exciting for students; this was part of a programme designed to 'accelerate language acquisition in hearing-impaired children'.

Hearing-impaired or deaf children have considerable difficulties in acquiring language and linguistic concepts, because the primary channel of reception – listening – is defective. The effort involved in interacting with language skills is so great, and the work proceeds so relatively slowly, that they have far less practice than other children. The system developed in California uses a series of syntactical patterns which are made real by the use of videotapes and other material. Each pattern is linked with sentences which illustrate them, and these sentences are written in every conceivable permutation. Advantages are that the whole arrangement remains under the direct control of the teacher (it is not 'automatic') and the interest level of the pupils is reported as high, thus enabling and encouraging a degree of practice which would be difficult without the aid of the computer system (Kreis, 1979).

DESIGNING FOR LANGUAGE DISABILITY

Recently, attention has turned to the potential of the uses of computers with those suffering from a specific language disability (or specific learning disability or dyslexia – the terms are used interchangeably). Here, the visual component of the computer may be of value, although such work is new. Kerchner and Kistinger (1984) state:

> The use of computers as word processors has become widespread among adults who do much writing. Word processing allows us to correct, edit, revise and manipulate text, making the process of changing first drafts to polished products less painful and less time-consuming. However, the use of computers by children for the same purposes has, thus far, been rare (p. 330).

These two authors used word processing facilities to approach the problems of the development of written expression with learning-disabled students, using as the assessment instrument the Test of Written Language (TOWL) (Hammill and Larsen, 1978). They report significant effects on the TOWL items of *thematic maturity*, *word usage* and *written language quotient*, and state that the 'positive results of this preliminary classroom study suggest that improved written language skills on the computer transfer to pencil-and-paper tasks'.

In considering another aspect of language function, spelling, Rieth,

Polsgrove and Eckert (1984) have devised a computer-based program to teach spelling, 'Spellmaster'. It was designed to allow maximum flexibility to teachers, and it:

> provides teachers with a pedagogically sound template from which they can build individualized student spelling lessons . . . . Field test data have yielded high teacher satisfaction ratings (p.64).

These researchers report that the data collected indicate that pupils with spelling problems have increased their scores on daily and weekly tests by an average of 50 per cent over the baseline.

DESIGNING FOR PERSONAL EXPRESSION

It could easily be argued that it is the difficulties in the development of perceptual capacities which are the greatest handicap of the disabled person during education, if not beyond. However, increasingly the new technology permits people to engage in activities which support the wider experiences of life, and therefore education. In 1978, Papert and Weir claimed that the computer can become an extension of the operator, who:

> can then do anything a computer can do, such as draw, compose music, gain access to information libraries, put text on permanent file, and so on (Papert & Weir, 1978).

They were also concerned with those aspects of the development of the individual which normally arise from exposure to and participation in an education environment. They say that the problem for disabled people, in terms of education, is:

> the restriction on their expressive power, so that such mental activity as does go on is trapped within the individual's own head (ibid.).

Papert and Weir pose fundamental questions, asking how linguistic structures may develop in people who have no experience of the production of speech. The same might be said of those who are deaf and thus have no experience of the receptive component of speech. They have examined the problems of the development of spatial thinking in people who cannot manipulate objects, working with Logo turtle geometry.

The Logo turtle is an excellent example of opening conceptual horizons for disabled people through the use of computers, both hardware and software. It was designed by Seymour Papert, and both able-bodied and disabled children are able to construct drawings with full consciousness of the processes which lead them to the end product (Papert, 1980). The turtle may be represented on a computer screen or be a real piece of equipment which moves about on the floor, under the control of the computer, which is in its turn under the control of the child. This is true whether the child is disabled or not. As the turtle moves, it traces a line, and thus the children are able to see very clearly the direct results of their instructions, and whether or not they have truly represented their own conceptions.

This relationship between the mental and physical world, which for the able-bodied may involve effortless linkages, is quite vital for the disabled individual. The results of handicap often force individuals into an unnaturally passive situation, preventing them from engaging in many interactions with the physical world or other people. Weir, Russell and Valente write (1982):

> A severe physical handicap imposes a dependent, passive role on its victim. The uncompromising way in which Logo places initiative and control in the hands of the users allows them to have a direct effect on their environment.

Possibly, it might be added, for the very first time.

The person with a visual handicap has a similar problem, in that there is a barrier to access to the real world, with a concomitant difficulty in terms of perception and conceptualization. However, the major area in which the blind individual has a communicative difficulty is in access to *information*.

DESIGNING TO PROVIDE INFORMATION

In the way that deafness is a major handicap in access to language because of the problems inherent in hearing and speaking, so blindness is a great obstacle in the way of access to information, because of the problems inherent in reading and writing. This will affect many aspects of everyday life, of course, but once again it is of crucial importance in the field of education.

For many years, the major route to information for many blind people has been through the use of braille. Braille text, however, is bulky, and, therefore, there are two versions of braille: grade 1, which is a transcription of the visual writing, letter by letter, and grade 2, which involves many contracted and short forms to save space. The transcription from text to braille (and vice versa) needs very skilled people; not only are they in short supply, but the process is also relatively slow.

An immediate use of the new information technology has been in this area of transcription, and there now exist a number of systems which enable a computer to carry out this task; an example is that developed at Warwick University (Gill, 1983). Use of these systems means that a braille version can be produced from a keyboard entry made by an ordinary typist with no knowledge of braille, as the machine does the conversion: thus is the manpower shortage solved. Many uses can be found for this development, and others will become apparent as time goes by and the possibilities become wider.

This sort of transcription system, while a great advance, still requires the human interface: in other words, an operator of some sort, and for large quantities of text this is still relatively time-consuming and expensive. The development of Optical Character Recognition equipment means that the visual information from the printed page can be 'read' by the machine and fed into the braille transcription system in digital form.

However, once the possibility of a machine 'reading' the printed, visual text is established, there is a further possibility of value to the blind person: this is transcription not into braille, but into sound. In other words, a machine that will literally 'read' out loud.

This type of system has been developed, and over several years has been refined to an extent that it is used by many blind people. The most widely known is the Kurzweil Reading Machine, which produces an auditory output in synthesized speech. The 'reading' end of the system is now sufficiently sophisticated that it can handle many different typefaces, and one of the most successful contexts for the system has been in libraries.

In many ways, the information storage, retrieval and manipulation functions of microcomputers have advantages for the blind person; but there is the practical problem that the feedback system, by which the operator may know what is fed into the computer or what is being read out, is visual – a television screen. This produces the anomaly that while the system may be potentially advantageous, the blind operator has problems in access to it. For the operator without sight, there are two channels of sensory input which can be utilized: auditory and tactile.

One might be forgiven for thinking that the simple exchange of visual material for auditory material – using a voice synthesiser to reproduce the sound of what is on the screen – would be a relatively easy and perfectly adequate solution to the blind person's difficulties. Certainly it is now, in computer terms, cheap and easy: the elements that compose speech may be synthesized into artificial sound, and there are many new ways to utilize this knowledge (for examples, see Poulton, 1983). However, it does involve the general difference between visual and auditory information: the auditory is presented in a unitary sequential manner. This means that the blind person relying on synthetic speech must receive the information in the order in which the machine speaks it. The sighted person does not do this (or at least, not all the time), for the visual representation of the information on the screen can be scanned in a variety of ways, and of course there is often much more information on the screen than simply the words of the text themselves.

Whether this matters depends on the type of information, but for a large number of processes which involve manipulation of the information – including word processing – inability to gain access to the non-text information, or inability to gain access to it fully, are major obstacles.

All the systems mentioned so far have involved a 'translation' from one information channel to another, usually from vision to hearing. Use of the tactile sense, while it involves a change of mode of reception, does not generally mean any difference in the structure of the information. This may be seen in such devices as the Optacon. This piece of equipment uses a miniature television camera to scan the visual material, and reproduces the shape of the letter in a tactile form on the blind user's index finger. Of course, this is a direct reproductive system; involving no translation, and requires the user to comprehend the shapes: it would be of no value to the blind person who was not familiar with the shapes of written letters. However, it has the advantage that it is able to follow any shape, not just print, and there is an adaptation which permits its use on television (and therefore computer) screens. Thus the blind operator may use it to obtain as accurate an analogy as possible of the visual material, as it is presented, including such items as layout, spacing, letter case, diagrams, etc.

It requires a high level of skill of the part of the user, however, and some find that the demands are very great indeed. It is a system to facilitate *reading*, in the sense of being a direct alternative to the skill of visual reading (rather than a 'translation', as has already been made clear) and therefore has its own framework of reading skills. These have been investigated, and ways found of increasing the reading skill by use of suitable teaching techniques, just as is the case with visual reading (Terzieff *et al.*, 1982).

DESIGNING FOR EMPLOYMENT

An area in which the new technology will create increasing opportunities of value to disabled people is that of employment: indeed this is a common prediction among people who have knowledge of the work situation of disabled people. The reasons are many and complex, but are well summarized by Scadden (1983). He lists nine factors which contribute to the increase of opportunity:

1. Computer and access technology can give disabled individuals equal access to most of the electronic information available today.
2. The increased use of computer technology in employment by 'knowledge' workers who work primarily with electronic information is increasing employment opportunities for disabled individuals.
3. The new economy appears to reflect a changing social order in which individuality and diversity are accepted more than in the past.
4. Specialization in employment is giving way to diversity of activity.
5. New industries are emerging as the backbone of the new economy.
6. The proliferation of computer-aided manufacturing which will utilize speech output terminals can provide many multi-handicapped individuals with tasks in industry.
7. The use of microcomputers and telecommunications systems will result in an accelerating number of industry and business personnel performing work tasks at home.
8. Small businesses are rapidly expanding in number, many of them 'electronic cottage industries'.
9. Education services are moving toward an emphasis on the use of computer technology.

There is much to achieve, however, in the social sphere as well as the technological. Pati (1983) reports a federal (USA) investigation of a large electronics manufacturer which revealed obstacles for the disabled. He lists them as:

1. A pre-employment medical examination that was not job-related.
2. Automatic exclusion of applicants on the basis of their handicap, even though there were plenty of jobs qualified persons with disabilities could do.
3. Discrimination against employees with invisible handicaps, such as obesity, color blindness, high blood pressure, allergies and varicose veins.

CONCLUSION

In an ideal world, every disabled person would have access to microcomputer technology which would enable them to carry out any function they required, whether that be direct or indirect. Examples of direct functions are:

- ☐ communicating, for those who cannot speak or hear
- ☐ reading, for those who cannot see
- ☐ moving about, for those who cannot walk.

Indirect functions include such things as:

- ☐ learning language
- ☐ acquiring education
- ☐ manipulating spatial concepts
- ☐ acquiring and maintaining employment.

In the early days of using information technology to assist the disabled, almost everything had to be designed and constructed specially. However, increasingly there are more and more useful facilities for handicapped people in equipment and programs intended for general use; some of this has arisen because it has been discovered (as every teacher has discovered) that systems which are advantageous to the disabled individual are usually of value to everyone else. Also, as technology becomes smaller and more portable, so the disabled can now carry with them devices which a few years ago would have required a trailer on the wheelchair! Thus the impact of the technology is increasingly on a rising number of facets of the individual's real life.

The debate as to whether it is best to invent special systems for disabled people or as far as possible modify normally-available material will in time diminish, probably to vanishing point. What must not be forgotten, whichever approach is adopted, are the people involved, both the disabled and those who support them. Hawkridge, Vincent and Hales in their book on information technology in the education of handicapped people (1985), say in their conclusion:

> The skill, dedication and patience of professionals who work with and help disabled people is legendary, and no technology can replace it. At best, the technology is a complement, often merely a supplement. By definition, no technology can provide a human caring environment. The human element is paramount.

One goal of work to assist disabled people could well be defined in terms of rendering it unnecessary to provide a special caring environment, in that they should be equipped to take their rightful place in the normal human environment. This raises serious questions of how such a goal may be achieved.

In time, there is no doubt that human beings will be able to remediate the problems engendered by disease, accident and congenital abnormality by reparative surgery, neurosurgery and therapy; indeed, in the field of prosthetics this is already beginning. So, one day we shall be able to give blind people new eyes, deaf people new ears, the mobility-handicapped new legs, and repair the malfunction of the tissue net in those with brain damage.

In the meantime we must use the tools we have to achieve as much as is possible, taking the steps that will lead us on the road to the final goal. During this process, which will take many years, we can increasingly use the advantages of constantly growing computer knowledge to improve and enhance the lives of those among us who have problems, of whatever dimension. This will not replace their human function, but enable them to express with ever-increasing ability that humanity which is masked by the barrier of inability to carry out certain tasks. Thus the person may fly free from the prison of the body, until we learn how to provide a new body.

REFERENCES

Beds and Bucks Advertiser (1983) Presentation joy for John. 1 June, 1983.

British Broadcasting Corporation (1984) *Horizon*, 27 March 1984.

Duncan, F.G. (1983) Substitutes for speech: Experience of applications of artificial voice production in micro-processor based systems for the disabled. Paper presented at EUROMICRO 83, Madrid.

Foulds, R.A. (1982) Applications of microcomputers in the education of the physically disabled child. *Exceptional Children, 49(2)*, 155–162.

Gill, J.M. (1982) Microcomputer aids for the blind. *Computer Education, 42*, 21.
Hales, G.W. (1978) Some problems associated with the higher education of deaf and hearing-impaired students in an open system. Unpublished PhD thesis, The Open University.
Hammill, D. and Larsen, S. (1978) *Test of Written Language*, rev. ed. Austin: Pro-Ed.
Hawkridge, D.G., Vincent, T. and Hales, G.W. (1985) *New Information Technology in the Education of Disabled Children and Adults*. Beckenham: Croom Helm.
Kerchner, L.B. and Kistinger, B. (1984) Language processing/word processing: Written expression, computers and learning disabled students. *Language Learning Quarterly, 7*, 329–335.
Kreis, M. (1979) Project Video Language: A successful experiment. *American Annals of the Deaf, 124*, 542–548.
Moses, J.F. (1983) Impressions of Eulenberg. *Rehabilitation/WORLD, 7(2)*, 14–19.
Papert, S.A. (1980) *Mindstorms: Children, Computers and Powerful Ideas*. Brighton: Harvester.
Papert, S.A. and Weir, S. (1978) Information prosthetics for the handicapped. Artificial Intelligence Memo 496: Institute of Technology Artificial Intelligence Laboratory, Cambridge, Mass.
Pati, G.C. (1983) A philosophical and cultural approach to high technology in rehabilitation. In L.G. Perlman and G.F. Austin (eds) *Technology and Rehabilitation of Disabled Persons in the Information Age*. Alexandria: National Rehabilitation Association.
Poulton, A.S. (1983) *Microcomputer Speech Synthesis and Recognition*. Wilmslow: Sigma Technical Press.
Rieth, H.J, Polsgrove, L. and Eckert, R. (1984) *Academic Therapy, 20 (1)*, 59–65.
Scadden, L.A. (1983) Employment and training opportunities for disabled persons in the computer age. In L.G. Perlman and G.F. Austin (eds) *Technology and Rehabilitation of Disabled Persons in the Information Age*. Alexandria: National Rehabilitation Association.
Terzieff, I., Stagg, V. and Ashcroft, S.C. Increasing reading rates with the Optacon: A pilot study. *Journal of Visual Impairment and Blindness, 76(1)*, 17–22.
Vanderheiden, G.C. (1983) Comparison of Apple, Epson and IBM microcomputers for applications in rehabilitation systems for persons with physical handicaps (revision C). Trace Research and Development Center for the Severely Communicatively Handicapped, University of Wisconsin.
Williams, M.B. (1981) What Emily Post never told me. *Communication Outlook, 3(2)*, 7.
Williams, M.B. (1982) Confessions of a closet technocrat. *Communication Outlook, 3(4)*, 12.
Weir, S., Russell, S.J. and Valente, J.A. (1982) Logo: An approach to educating disabled children. *Byte, 7(9)*, 342–360.

THE COMPUTER IN EDUCATION: A FORCE FOR CHANGE?

Geoffrey Underwood
and
Jean D. M. Underwood

Twenty years ago Patrick Suppes (1966) argued that developments in educational technology, and specifically in computer usage, would change the face of education in a very short space of time. He based this prophecy on the unique capabilities of the computer to be used interactively, to present materials in novel ways not easily available through other media, and on the flexibility of the machine in adapting to different learning and teaching styles. Some level of computing capability is now available in the vast majority of British schools, but education appears to have changed little.

There are many reasons for the non-fulfillment of Suppes's prophecy. Richard Clark (1984) points out that we are continually searching for new technologies that will produce more learning outcomes than 'older' media, and that the computer is seen as a particularly powerful technology in this sense. Clark questions the role of the computer in education, however, arguing that it is very easy to mistake sophisticated technology for sophisticated learning, assuming productive outcomes when students communicate with computers.

There are indeed many examples of complex technology being used to achieve low-level educational goals. The National Science Foundation of America's Time-Shared Interactive Computer Controlled Information Television project, TICCIT for short (Bunderson, 1974), and the current development of distance learning are two cases in point. In these projects the machine is used for the largely didactic transmission of knowledge from 'teacher' to 'learner,' and subsequent testing of the learner's acquired knowledge base.

Chambers and Sprecher (1980), on a less negative note, suggest that the single most critical issue today in computer based learning

(CBL) is the development and sharing of quality software. Indeed, the lack of good software is frequently cited by teachers for their reluctance to incorporate CBL techniques into their own classroom practice.

Now that many, although not all, of the initial technical and economic problems of computers in education have been resolved with the advent of cheaper and increasingly reliable microcomputer systems plus substantial funding from national government, this emphasis on software is understandable. However, the questions arising out of the spread of new technologies into the classroom are not just issues of how to put traditional lessons onto the machine in an efficient and economical manner, although much of the current research is directed towards that end. There are more fundamental issues, as Clark suggests, which spring from the role of the computer as a large and efficient store and processor of information and as a powerful tool to think with. Although the classroom computer can be used as a flexible tool for activities as diverse as text processing and simulation modelling, there is a risk of it being used as a 'teaching machine' in a very conventional classroom setting.

A growing number of teachers are using the tool-like capabilities of the computer to show things, to say things or to provoke thought. However, it is often the tool user who has most fun and who gains most value from the encounter with the machine; unfortunately, in many cases that tool user is the teacher. Ogborn (1985) argues, therefore, that developments in computer-based learning (CBL) should emphasize the tool-like uses of the computer, but that the learner should take the active role of operator. In encouraging children to use the computer as a tool, we are mimicking out-of-school uses of the computer, with these children participating in activities relevant to such environments as the electronic office, and engaging in an exploration of knowledge in a way that is akin to current research in artificial intelligence.

We would suggest, therefore, that one of the major issues in educational computing is the definition of the goals of classroom computer use. Amarel (1984) argues that questions which are being asked about the educational benefits of chip technology have resulted in the computer becoming a focus and catalyst for change in education, because they mirror the diversity of views about the proper goals and functions of schooling. The concept of the computer as a tool opens up new directions in education, challenging the roles of both teacher and child and questioning our definition of 'worthwhile' knowledge and skills.

Our intention here is to consider three issues in CBL implementation, the resolution of which will influence the nature of and the prospects for widespread and worthwhile educational computer use. The first section reviews the methods by which we should evaluate the likely benefits of CBL, using a sample of recent studies. The second section considers the ways in which CBL could be used to enhance the development of children's minds, and the third section speculates on the possible changes in social interaction which might accompany the successful implementation of CBL.

EVALUATING THE EDUCATIONAL IMPACT OF CBL

Despite the general enthusiasm for the introduction of computers into the classroom, or perhaps because of it, educators will need to be convinced that this change in teaching practice will produce an educational profit. There have been too many educational vogues for us to be surprised when teachers express scepticism about this latest innovation. Whatever happened, for example, to the initial teaching alphabet or programmed instruction? With a history of unfulfilled promises we need to demonstrate that CBL offers benefits over traditional classroom practice. To do this would be no mean feat, for several factors are working against us, including the generally poor quality of available software, and a misplaced humanistic reaction to the concept of evaluation. Clark's (1984) assertions that the evidence, to date, is highly controversial and that supposed learning from CBL can be attributed to uncontrolled effects of content, instructional method and novelty, demand a more rigorous approach to evaluation. Even if we succeed in demonstrating that a CBL package can offer improved learning or better understanding there will be a group of diehards arguing that statistical, comparative evaluation is inappropriate, and that teacher-friendliness or teacher-perceived usability is the only basis for introducing anything new into the classroom. This section of the discussion reviews a sample of the current evaluations, concluding that CBL can indeed make teachers more effective, and can also make their job more enjoyable. It also challenges the view that the user of the product is in the best position to evaluate its usefulness.

Motivating effects of CBL

Does CBL help? A number of recent investigations, using a variety of software packages but mainly within the domain of mathematics, have suggested that it can. One of the most extensive evaluative studies was reported by the ITMA Group (Ridgway *et al.*, 1984), using programs which were found to avoid repetitive drill-and-practice and which developed important activities in the mathematics classroom previously identified as neglected by the Cockcroft Report (1982) on mathematical education.

There was an increase in activities such as the discussion of hypotheses and problem-solving, but Cockroft's missing 'investigational work' was not facilitated even though investigative software was available. Ridgway *et al.* concluded that such packages were not used because teachers could find no way of fitting them into the existing curriculum. There is a message here about the relationship between the development of children's minds and the design of the ideal curriculum. If a desirable educational procedure cannot be integrated into the curriculum, then doubt is cast upon existing lesson aims and content. Perhaps this will be one of the consequences of CBL: it could allow us to achieve educational goals currently excluded from the curriculum.

Secondary school teachers in Ridgway's study were free to select software of their choice from a library of about 90 programs, and their opinions were determined by asking them to rate each program on a scale from 'I would never use this program again' to 'I would use this program again even if the computer had to be booked a week in advance and carried up two flights of stairs before the lesson.' Ridgway *et al.* found selective enthusiasm on the part of the teachers – six of the programs proved to be particularly popular, and these had no identifiable features in common other than general acclaim from the schoolchildren. The programs covered a range of mathematical activities including algebraic notation, coordinate systems, arithmetic, graph interpretation, and coordinate/vector relationships, and children responded with comments such as 'I wish all maths lessons were like that' and 'I like doing this instead of maths'! Programs were sometimes used by teachers as rewards for the completion of non-CBL activities, and whereas this is clearly a poor use of the programs, it does serve to indicate their popularity in the classroom. Part of the reason for the success must lie with the video games format of the most popular programs. By making use of children's competitiveness, CBL can offer enormous educational benefits. As Ridgway *et al.* conclude, if this format can increase children's enthusiasm for mathematics, then it must be worthwhile. The availability of motivating programs is in itself not sufficient to guarantee their optimal use, however. Allowing children to use programs as a reward for completing other activities could act to reduce CBL to the equivalent of arcade stimulation.

The study reported by Ridgway *et al.* provides both excellent support for the introduction of CBL and a number of cautionary notes about potential misuse. It does not demonstrate that the positive benefits derived from computer-based lessons could only be achieved this way: it is entirely possible that the children's enthusiasm for mathematics could have have developed using traditional classroom practices. The fact that CBL can change affective responses to mathematics is in itself very encouraging, but the study does not provide evidence of the specifically *computer*-based nature of this change and, even more importantly, it does not attempt to evaluate the effects of CBL upon the children's understanding of mathematical concepts.

Cognitive development

Having established that children like their mathematics to be presented as part of a computer game, we turn now to the evaluation of CBL as an aid to the development of their minds. We start with a study which observed the effects of CBL upon classroom interaction. Chatterton (1985) used a formal notation of interaction analysis which allows an observer to estimate the level of guidance given by one classroom participant to another, and this small investigation has the advantage of allowing us to compare traditional lessons against CBL lessons. The use of CBL was associated with an increase in child – child dialogue and a reduction in the amount of teacher-based explanations. Increases were also reported for individual teacher – child interactions and for the numbers of groups of children engaged in problem-solving discussions. Moreover, the nature of the questions asked by children also changed in the CBL lessons: specific questions of the 'what temperature do we use?' type were replaced by more general, predictive questions of the 'what happens if we increase the temperature?' type. Chatterton found that children in the CBL lessons were more likely to question the reasons underlying the facts generated by the computer model, and more likely to discuss hypotheses about the relationships between the facts. Again, we cannot be sure that Chatterton's very desirable changes could be achieved only with the use of CBL, but this does not detract from the improvements that were actually observed. CBL appears to make these improvements readily obtainable.

The interactions between children and teachers change when computers are introduced into the classroom, and one of the more interesting changes reported by Chatterton is towards more hypothesis-based questions. This result was also reported in J.D.M. Underwood's (1986) study of the effects of two data base programs upon questioning strategies in primary school children. This experiment was concerned with the development of classificatory abilities in 10-year-old children – abilities identified by Bruner, Goodnow and Austin (1956) as being necessary for perceptual discrimination, for ordering the relationships between objects and events, and for efficient encoding and retrieval, among other cognitive acitivities. An efficient categorizor might be considered to be a perceptually alert and efficient learner.

Underwood's (1986) study is described in some detail here, because it demonstrates the power of the empirical evaluation of CBL. Groups of children were first matched for reading ability, because reading skill has also been shown to be associated with classificatory skill. The groups were also matched on IQ and on a pre-test categorization task. This pre-test required the children to ask questions in order to identify which one of 24 pictures the experimenter 'had in mind.' The pictures were scaled along dimensions of shape, size and number, and invited set-membership questions to which the answer would be 'yes' or 'no'. All the subjects then engaged in a three-week class project in which the main task was to classify a range of common cheeses. The experimenter was not involved in this project. Half of the children used commercially available data-bases on a classroom computer, and the rest simulated the use of the data-bases using index cards. So that each group of children had similar computer experience, the control group completed a series of CBL mathematical investigations. After the cheeses had been classified, the categorization task with the 24 pictures was repeated.

Two results are of importance here. First, the CBL data-base children showed an improvement in the number of questions asked about the pictures – they required fewer questions in order to identify the target picture – and second, they asked more constraining questions and fewer specific questions. A specific question in this context would be 'Is it the picture of the three small apples?' and a constraining question would be 'Do you have one in mind?' The index card data-base group showed no improvement in classificatory ability. Although the experiment does not attempt to show any long term benefits of CBL, it does demonstrate the gains that can be made in teaching children how to think about the organization of their perceptual world. The experiment also emphasizes the importance of matched groups, and of pre and post-test comparisons. The improvement of the CBL group is only brought into relief by the lack of improvement of the index card group.

The use of control groups does not seem to be a popular procedure in educational research; neither does the use of performance measures before and after the CBL treatment. Underwood's (1986) classification experiment is unusual in this respect, but one other example can be mentioned. Nicolson, Bowen and Nicolson (1984) looked at CBL methods with primary school children, observing the effects of teaching the elements of arithmetic to eight-year-olds. A control group was taught using traditional methods, and an experimental group, matched on initial arithmetical ability on the basis of a pencil and paper test, was given a CBL arithmetic package instead of one of their other mathematics lessons each week for six weeks. Both groups improved in comparison with their pre-test scores, but the CBL group improved differentially. An analysis of the errors made in the post-test indicated that the CBL group were making fewer arithmetical rule errors than the control group, but there was no change in the number of computational errors. There are a number of possible reasons for the improvements shown by the CBL group in this experiment, including those attributable to the advantages of small group discussion, those attributable to the CBL group regarding themselves as the "special" group (cf. the Hawthorne effect), and those attributable to the associated motivating effects of CBL in teaching mathematics (see Ridgway *et al.*, 1984). However, the advantage of CBL procedures was again demonstrated, and in this experiment the advantage was seen by comparison with a control group.

Empirical evaluation is necessary

If the experimental method in the agricultural tradition is so rarely used in educational research into the benefits of CBL, then either the alternative method of evaluation has enormous advantages, or educators are already convinced of these benefits and need no statistical data to support their beliefs. The enormous variation in the extent of CBL use is sufficient to suggest that the evaluations have yet to convince everyone, so what are the alternative methods of evaluation that are so compulsive? Hartley (1984) has considered the power of alternative evaluation procedures, comparing empirical with 'illuminative' or descriptive methods in the assessment of the use of a mathematical modelling package by undergraduates. The opinions of the course lecturer provided evaluative comment on the difficulty of integrating the package, and on his students' positive reactions to its introduction. The lecturer felt that the exercise had 'yielded significant insight into better ways of using the package to achieve his objectives' (p. 75). In addition, Hartley used a post-test with randomly matched pairs of CBL and non-CBL students, finding that the CBL group performed at a *lower* level than the control group. The computer package seems to have impeded understanding in this study, but as there was no matching of students' abilities prior to CBL exposure, we cannot be sure that the difference between the groups was not present from the start. However, this study is interesting in that it provides one of the few results which goes against the generally positive view of CBL advanced here. It requires that caution be used, but, although the experiment does not possess the features of our ideal evaluation design, caution alone is probably sufficient. More importantly, the experiment raises the issue of alternative methods of evaluation (if evaluation X doesn't work, try Y?), and again demonstrates the importance of thorough experimental design. A poor experiment is arguably of less use than no experiment.

A feeling sometimes voiced in defence of empiricism is that the critics are unhappy with experiments because they are unable to conduct them correctly. There certainly exist plenty of examples of poorly controlled experiments which give uncertain conclusions, whose authors are amongst the critics of empiricism, but we surely cannot reject a methodology on the basis that someone, somewhere is unable to use it cleanly. Only criticisms of well-designed experiments can be used in the evaluation of evaluation, and some of these criticisms will reject empiricism on a priori grounds of the inappropriate nature of experiments in the classroom. Perhaps the critics will say that the educational development of a child's mind cannot be assessed under the conditions necessary for experimental testing, or that it is the development of the individual which matters, rather than changes in group averages. We disagree, partly because we have shown that it is possible to conduct informative classroom experiments; but let us first consider the position taken by the critics.

The trend towards an empirical evaluation of CBL was considered and found wanting by Kidd and Holmes (1984). The main problem with empiricism, for these authors, is that it can only evaluate a treatment under a specific set of conditions. This is, we would contrarily reply, the strength of empiricism: the conditions of testing are well-defined and do not (or at least, should not) vary along unknown dimensions. However, this strength is sometimes mistakenly seen as a weakness, in that the results can apply only under the conditions of the test. This is true, of course, of the evaluation of any innovation, from the introduction of a new procedure for teaching a second language, to the comparison of blackboards and overhead projectors. The treatments certainly depend upon a minimal skill on the part of the teacher, and effectiveness can only be demonstrated to be possible under certain circumstances. For innovations to be effective, they must be used appropriately: part of the empirical test should be to determine what constitutes 'appropriate' use. Even the most accepted and useful of educational innovations are unlikely to be foolproof. Part of the specification which accompanies a CBL package, therefore, should contain guidelines for its use, and these guidelines should, wherever possible, refer to the established effectiveness of a particular use.

Teachers, like the rest of us, do not always use innovations appropriately. Char *et al.* (1983) monitored the use of one navigational simulation (RESCUE MISSION), finding that those teachers in sympathy with the designer's aims used the program to supplement the mathematical curriculum, and identified children's difficulties with the program as being mathematical in origin. In contrast, those teachers who saw the program as a specific aid to teaching navigation tended to relegate the program to non-lesson time, and suggested that the children had difficulties because they had never been on a boat! Equally, guidelines do not always facilitate optimal program use. The simulation MARY ROSE, sponsored by the government's Microelectronics in Education Programme, potentially encourages the development of spatial concepts and is based on a real-life historical event, the sinking of the Tudor flagship. Although the program guidelines are clear, they are subverted by the teacher-designed back-up material which accompanies the pack. Totally ignoring the mathematical potential, the majority of the worksheets involve the collection and recording of historical facts of the time, such as 'Draw or trace a Tudor mansion.' This simulation seems a very expensive prompt to a tracing lesson!

With any educational aid – a CBL package or a textbook – the potential for use can be lost outside a specific set of conditions for use. If an experiment demonstrates that certain conditions of testing are associated with educational effectiveness, then those are the *established* conditions of effective use. Other conditions may also be effective, and we would all be loath to stifle teacher or pupil creativity, but it is difficult to see why Kidd and Holmes reject the usefulness of knowing *when* an educational practice will be effective.

As an alternative to empiricism, Kidd and Holmes recommend

evaluation based upon informed opinion. There are several problems here. First, the informed opinion often comes from the committed user. What reply could we expect from Seymour Papert if we asked him whether LOGO was an effective educational tool? Second, if we rely upon the personal judgements of new users, we run the risk of insufficient time and resources being allocated to the package for the potential to be appreciated. Finally, the classroom teacher will not necessarily have the training required for evaluation. We should not underestimate the critical powers of the practitioner, but Preece and Jones (1985) have found that teachers are inconsistent judges. Even after a short training course on the evaluation of software, teachers were uncritical, and their summaries did not appear to be related to their evaluations of component features. They offered very few constructive criticisms of how the programs could be improved.

Clearly, teachers need to be able to evaluate any materials or methods which they are to use in a classroom, but it may be unwise to restrict the evaluation to the 'informal opinions' advocated by Kidd and Holmes. The teachers may actually welcome information about the likely effectiveness of a CBL package and about the best conditions for observing this effectiveness. The empirical methods of the classroom psychologist are ideally suited to providing these evaluations.

THE IMPACT OF THE COMPUTER ON THE COGNITIVE CLASSROOM

One of the greatest problems we face in trying to analyse the impact of computers on education is the very versatility of the machine. It can be used to great effect as a calculator, a 'teaching' machine, a processor of complex information and a creator of microworlds. In essence this means that the computer is a tool which can support the full spectrum of educational philosophies: for example, acting as a tutor for those who believe we should return to a basic skills curriculum or as a key factor in stimulating the dynamic process of writing.

The debate concerning the most profitable ways in which we can use this costly and, in some schools, scarce resource can simplistically be divided into that of the computer as 'teacher' (the sub-skills tutoring/practice approach) versus the computer as tool. At the moment, it is apparent that the practice of basic skills dominates the educational use of the computer, a fact which leads Chandler (1984) to suggest that this machine has made it possible for educational practice to take a giant step backwards into the nineteenth century! Rubin (1983) echoes the complaint that there is a concentration on the development of low-level skills in reading and language. He argues that this is due in part to the nature of the computer, which encourages a mechanistic, detail-oriented education, focusing on 'correct' or unambiguous answers; but also to the ease with which workcards and books, concentrating on sub-skills, are converted into programs.

Amarel (1984), however, considers that the superiority of practice programs is compelling in comparison with the workbooks or sheets that they replace. In well-designed programs children can receive quick and accurate feedback, progressing at flexible rates, without publication of errors. The educational computing journals all too often quote from interviews with children, whose main motivation for wanting to use the computer is the patient, kind and generally non-condemning attitude it presents to their follies.

The practice of such skills is not inherently wrong – indeed practice is vital if skills are to reach the level of automaticity necessary to allow the individual to focus attention on higher-level problems (Underwood and Underwood, 1986). If sub-skills practice is all there is to education then it is sadly impoverished. Basic sub-skills practice is essential, and the use of computer can not only make practice less tedious for all concerned (by building it into a games-format package), but can also make practice less the focus of the activity (by integrating practice with the development of other skills). One of the great benefits of the high capacity machines now available is the facility of integrating activities at different cognitive levels. The same program may require sub-skills arithmetic practice and three-dimensional navigational mapping, with overall success in the simulation game depending upon success at both levels. For example, relatively simple programs such as Richard Phillips's ERGO or Anita Straker's GUSINTER deal with an understanding of multiples, but they also require the child to use that knowledge in such a way as to solve a more general problem. The user is engaged in sub-skills practice through goal-directed activity. Drill-and-practice is necessary for the development of automatized skills, but it does not have to deal with children as automatons.

It is deeply worrying that Amarel (1984) should find that drill-and-practice programs are offered to low achieving or disadvantaged children in the belief that they will benefit most from these 'highly structured, non-judgemental, infinitely patient environments?, while their advantaged peers enter the expansive world of the computer as a tool. This differential use is seen as encouraging a questioning autonomy in one group of learners, while their less favoured peers are caught in an environment which sets limits on independent thought, curtailing the exercise of options and the opportunity to reflect. It also provides one group with the skills and insights into the uses to which a computer may be put outside the classroom, while the drill-and-practice class sees the computer as an electronic worksheet, a concept of little environmental value.

The two-tier society envisaged here is unnecessary; the use of creative, adaptive software need not be confined to an élite. In the area of children's writing, it has been established that children will concentrate on low-level skills of text production, such as spelling and handwriting, when the task at hand becomes too complex for them to handle successfully (Bruce *et al.*, 1982). Software which reduces children's needs to concentrate on such low-level skills provides the opportunity for children to direct their attention to higher-level skills such as the logical flow of narrative. Gary, a 10 year-old child, whom we introduced to the word processor, increased his average writing output by well over 500 per cent, on discovering that with the processor and most importantly the printer, he could actually read what he had written!

Rubin's (1983) STORY MAKER is a tool which allows children to create stories by selecting from a set of story segments presented by the computer in a structured tree. In using this program, children never revert to low-level skills, because those details are resolved by the program. His description of story production in a mixed class of readers and non-readers is particularly promising. Here two children who could read were creating a science fiction storybook. As they made each choice in story direction they read their work out to a group of non-readers. Soon these non-readers were contributing to the storyline and were involved in an enjoyable creation of text even before acquiring the skills to read their own stories. Rubin offers no data on the motivational impact of such a learning experience but it is difficult to envisage that it was not significant.

Cognitive development

We have already argued that there is a need to develop rigorous evaluative techniques in order to assess the impact of the computer on individual understanding and attitude. We cannot simply assume that because it looks right then worthwhile educational events are taking place. What evidence is there that the use of the computer can affect cognitive development?

Hoyles, Sutherland and Evans (1986) provide evidence of cognitive change from their case studies of secondary school children working with turtle graphics aspects of LOGO. In this project pupils were encouraged to set their own goals with minimum teacher intervention, provoking highly motivated children to achieve very sophisticated goals which inevitably involved complex mathematical ideas. They defined these goals on two dimensions: the degree of abstraction (abstract mathematical versus picture) and level of definition (well-defined versus loosely-defined). They found that the selection of one of these four main types of goals influenced the learning outcomes of the children in their study. They concluded that children who selected well-defined abstract goals, in this case the production of a complex star shape, were more likely to see the modularity in the goal, that is they were both able to break down a problem into its component parts and to put the pieces together to produce a final solution, and they were highly motivated to extend their work. An important part of this goal resolution was the activity of 'making sense of' – a term they use to represent the process by which a concrete image is used to think with in order to make sense of the world, the equivalent of Papert's (1981) gear wheels.

The work of Hoyles, Sutherland and Evans (1986), and of other workers, highlights the different cognitive styles children bring to their work on the computer. The research in this area is important because the illumination of the thinking process emphasises to teachers the multiple ways in which a problem can be approached. In the final analysis the achievement of a goal tells us far less about the cognitive progress of students than does the path by which they reach that goal. Hoyles, Sutherland and Evans (1986) discuss two children who produced a series of interesting visual images on the screen, but who showed very little understanding of LOGO principles at the end of a year's work. The children had borrowed a procedure for a rotated pattern from neighbours early in the project, and had achieved their various visual effects by simple changes of the length and angle. They had developed a recipe for success without any deeper understanding. Any attempts to be more innovative resulted in initially less satisfactory screen outcomes, and they therefore gave up at any pretence at exploration. Hoyles and his co-workers suggest that this operation at a syntactic level with an absence of process involvement can lead to boredom and a lack of self-confidence. As discussed more fully elsewhere (Underwood, 1985), it also poses a problem for teachers who are used to assessing outcomes rather than process.

Further evidence of the impact of the computer upon cognition is offered by Smith (1986). He found that engineering students, when problem-solving with the aid of a computer simulation, were more likely to think about the engineering realities of the problem than when working with pencil and paper. In the latter case students reduced the problem to a mathematical exercise.

The effect of CBL upon the development of children's classificatory ability in J.D.M. Underwood's (1986) study has already been described in some detail. In a standard pre/post-test evaluation she found that children working with data bases on the computer performed significantly better on the simple classification post-task than did a control group. In particular, the computer children showed greater facility in the use of constraining or superordinate questions. The improved test performance by the computer groups was independent of age and ability in this experiment. Children in the computer groups did not out-perform the non-computer users on a test to assess the recall of factual information acquired during the period of the project. In this situation the learning of factual information was related to measures of verbal and non-verbal ability. The computer in this project was used to highlight the processes of organizing data and this appears to be what the children have learnt.

The use of information-handling packages will not guarantee educationally beneficial results. In quizzing a data base, it is suggested that even young children can begin to ask 'good' questions and be introduced to an hypothesis-testing strategy to learning, but the questions children ask are often seeking confirmatory evidence. Thus a group of infants quizzing a data base called 'Myself' asked the computer to find all the children with blue eyes and blond hair. At the age of six years, these children can be forgiven for not asking for those cases which would test their hypothesis, particularly as the limited screen information presented after a sorting operation with FACTFILE (the data base in use), encourages hypothesis-confirmation rather than hypothesis-testing through disconfirmation. Children also show this tendency to look for supportive rather than disconfirmatory evidence when playing explorative games such as MARY ROSE or PHILOSOPHER'S QUEST.

Chandler points out that although adults also show this tendency towards hypothesis-confirmation (Wason and Johnson-Laird, 1972), as teachers we should be challenging rather than reinforcing these problem-solving strategies. The pragmatics of achieving such a goal are far from clear, however readily we may endorse it. We have argued elsewhere (Underwood, 1985) that, as with much of the work connected with computers, it may be easier for children, rather than teachers, to develop these 'new' ways of thinking. Our experience suggests that teachers themselves need practice in more explorative approaches to learning. They need to be weaned away from an emphasis on end-products, towards an appreciation of the pathways towards a goal. This may prove easier for the student than the teacher. Not only have children less to unlearn when they approach new learning environments, but as many CBL packages are based on a games format, they may be more readily accepted by children steeped in the culture of arcade and home computer games than by teachers, whose experience of these leisure activities is limited.

THE IMPACT OF THE COMPUTER ON THE SOCIAL CLASSROOM

In their study of teachers' attitudes to the introduction of educational computer usage, Johnson *et al.* (1981) found that one of the deepest concerns was that of the impact of the computer on the social equilibrium of the classroom. Teachers were particularly confused as to their own changing role following the technological innovation. The identification of such a concern is not surprising. Many pioneers in this field argue that the use of the computer can, and possibly should, influence the roles of both learner and teacher. The most revolutionary changes are of course predicted by Seymour Papert (1981), in the event of the successful implementation of LOGO microworlds into the classroom. Less dramatic, but nonetheless significant shifts in power and group dynamics are inherent in the ITMA Group's emphasis on the teacher as 'co-learner' (Ridgway *et al.*, 1984).

Teachers in studies by Lichtman (1979) and J.D.M. Underwood (unpublished manuscript) did not exhibit similar role-fears. The findings from these studies are not necessarily contradictory, however. In the latter study, conducted over a three-year period and sampling the views of 306 teachers across the school spectrum, those teachers with generally positive attitudes to the computer, over 80 per cent of the sample, recognized that changes in social structure would take place. In particular they expected greater autonomy for the learner, the corollary of which must be a redefinition of the teacher's role. The disparity with the findings of Johnson *et al.* lies in the fact that these changes were perceived as an advantage of CBL rather than a disturbing development. For those who were generally negative, there was again little fear expressed, but these teachers seemed to operate on a principle of cognitive dissonance, concluding that a machine of such little educational worth would have minimal impact on the classroom. For these teachers this may well prove a valid statement, for they seem to like to keep the computer in a cupboard with the door firmly shut.

Amarel (1984) argues, however, that, whatever the assumptions or attitudes of the educational fraternity are concerning new social structures in the technological classroom, the probability of change is slim. Classrooms are intrinsically stable settings with well-established cultures, social dynamics and work-related agendas, all seated in the established curriculum. Arguing from Roger Barker's (1968) tenet that 'settings have plans for their inhabitants', she suggests that this stable setting, which has resisted the influence of numerous innovations, is unlikely to be re-organized by even such a powerful tool as the computer. The demand characteristics of the classroom will continue to be paramount in the prediction of people's behaviour.

Despite strong pro-computer attitudes, the criterion for the selection of all the teachers involved in the PLATO project, the study found considerable disparity in the level of impact and change that teachers permitted to take place and Amarel emphasizes the powerful effect teachers had on the process and, ultimately, the outcome of the implementation. This influence was shown in the disparity between mean pupil-time on-line, which varied from seven to 70 hours. In *laissez-faire* classrooms computer 'jocks' commandeered the terminals, to the detriment of less forthcoming children, who were frequently, but not always, female.

The rise of pupil expertise

Teachers also hold strong views on permissible student interaction with the computer. Amarel (1984) reported that most teachers were unhappy when computer contact took place outside the formal setting of the organized lesson, but a small band of teachers saw such contact as providing new and welcome educational opportunities. For example, one teacher removed several pupils from the machine on the grounds that they had completed their alloted task and because it was no job of theirs to tutor other students; that is, he or she was unhappy about pupils stepping outside their role as learner and infringing on the role of the teacher. This act denies the very experience that workers such as Papert (1981) and Turkle (1984) consider to be one of the great strengths of CBL, the growth of student autonomy. A second teacher, however, actively set up self-help groups, each with its own child leader to co-ordinate the group's work.

The reaction of teachers to pupils' growing computer expertise is also illuminating. One teacher closed down the system because pupils were hacking into the programs rather than 'doing maths'. The contrast came from a teacher who encouraged pupils to gain computer know-how by watching and experimenting, in order that pupils should gain increasing levels of independence.

This fear of pupil-expertise threatening the teacher's role of 'repository of all worthwhile knowledge' is very real. Daiute (1985) suggests that although children and teachers can collaboratively achieve productive outcomes in writing classes, recent work suggests that the use of the computer makes the writing process more public for children, and they are more willing to welcome their teacher as a partner rather than as a manager. The benefits of such collaboration are that the 'the teacher who is willing to take some risks in writing at the computer with students has a chance to learn as well'. Note that Daiute recognizes that there is a risk: a risk that teachers with low self-esteem may feel unwilling to take, as is confirmed by the PLATO evaluation (Amarel, 1984).

Our own experience, in monitoring the introduction of LOGO turtle graphics into a primary school, also provides some evidence of the strength and influence of established role patterns. This small pilot study involved children from both the lower and upper primary age-range, who worked in small groups with both the floor and screen turtle. By the end of the sixth week, much to the delight of the teacher, it was clear that each group had developed acknowledged experts to whom the children would refer when in difficulties. To our surprise both a male and female expert had emerged in each of the groups but, disturbingly, although the children consulted either expert in the lower school groups, in the upper school, by the age of 10, the female experts were largely ignored. Computers had become a male knowledge-domain supporting already established social structures rather than creating new patterns of behaviour.

The facts so far suggest that the computer is having very little impact on the social structure of the classroom. The teacher is still the pivotal point of any social dynamics, and if he or she has a didactic teaching style this is how the computer will be used; equally, teachers wishing to give their pupils greater autonomy will find a ready helper in the computer. Is there no evidence then of social change in the classroom following the introduction of one or more computers? A number of workers have noticed that on introducing a computer into the classroom the interactions around the microcomputers are different from those associated with other activities (Ridgway *et al.*, 1984; Chatterton, 1985). Contrary to the image of child automatons tied to their robot master, there appears to be more social interaction between students and between students and teachers in the technological classroom.

The rise of collaborative work

Hawkins *et al.* (1983) concluded that the advent of a computer into the school setting can disrupt the social organization of work in interesting ways. Interviews with children, as young as eight years old, reveal that they have firmly established opinions of when collaborative work, as opposed to individual work, is appropriate. In the former case 'fun' activities such as drama are collaborative but 'serious' subjects such as mathematics have to be completed by the individual. When children were using the computer, however, they observed more peer group collaboration, more solicitation of help from other students, and more 'dropping in' to make comments and suggestions when the children were programming, compared to work on other tasks off the computer, which were designed by the teacher to be interactive. Turkle (1984) supplies support for these experimental findings.

Chatterton's (1985) observational study of science lessons and the ITMA studies of mathematics (Ridgway *et al.*, 1984), confirm that these benefits extend beyond programming into more traditional teaching areas. In each case CBL brought about a qualitatively improved learning environment in which pupils were more inclined to discuss the issues to hand, developing skills in questioning, hypothesis-testing and collaborative approaches to work.

Sheingold, Hawkins and Char (1984) suggest that the potential for emphasizing the types of learning interactions we have been discussing raises issues for both students and teachers. They argue that students do not necessarily have well-developed collaborative skills, nor are they very good at using human resources other than teachers in their work. Their example of two little girls LOGO programming, in which the dominant child assigns the role of 'typist' to her friend and of 'thinkist' to herself, is a salutory statement on the realities of classroom interaction, and is in stark contrast to the studies discussed above. The more creative interactions are often teacher initiated by the careful management of human resources. Although such a role is not confined to activities with the computer, evidence that more group work takes place when a computer is part of the lesson plan (Chatterton, 1985) leads us to suggest that this is one of the teacher's skills which will become increasingly important in the technological classroom.

This conclusion leads us full circle to our initial query about the role of the teacher once a computer is introduced into the classroom, and the consensus appears to be that the teacher should act more as a facilitator of the learning process. This involves managing the learning environment, probing the children's understanding of material, providing key intervention when a child is in difficulty, and all this is set in an environment where the teacher is in partnership with the learner; a partnership which may also extend to the teacher adopting the role of learner alongside his or her pupils (Ridgway *et al.*, 1984; Daiute, 1985).

Good teachers have done this all along, but the move to more collaborative work stimulated by CBL packages thrusts such an approach on a wider range of teachers. Whether the work is in problem-solving or in creative activities such as writing a newspaper, it emphasizes the processes by which goals are achieved rather than the final reaching of the goal. This places yet another strain on the teacher, which is how to evaluate the learning outcomes from such interactions (Sheingold, Hawkins and Char, 1984; Underwood, 1985). This is turn begs the question whether teacher trainers are passing on appropriate skills to their charges.

THE EDUCATIONAL IMPACT OF CLASSROOM COMPUTERS – AN ILLUSTRATION AND SOME CONCLUSIONS

The classroom computer has the potential for creating an enormous change in the way we encourage the development of children's minds. By way of indicating some of the changes that can be achieved, this chapter closes with an illustration from a computer-based 'lesson' using a binary classification program called SEEK. This is one of the programs used in Underwood's (1986) experiment, reported earlier. The illustration takes the form of a transcript from the classroom interaction – it is a short play in three acts. There are five actors – three children (C1, C2 and C3), one teacher (T) and the computer (CBL). The play might be called *Changes in the Locus of Educational Control* given that the events of particular interest involve the children, teacher and computer taking control of the interactions at different points. Just as Papert found that LOGO enables children to take control of their microworlds, this program encourages these eight-year-old children to question answers suggested by the computer, to 'teach' the computer to describe a new datum entry and, perhaps most importantly, helps the children discover the importance of asking good questions.

The lesson is part of a botanical project on the identification of trees. This part of the transcript concerns the uses of leaves as identifiers.

ACT 1: *With the Computer (Computer as authorative source).*

CBL + C1: Is the leaf just one simple leaf on a stalk? *(C1 acts as computer operator throughout, reading aloud screen instructions, and entering responses via the keyboard.)*
C2: What does 'simple leaf' mean?
C1: It means all in one piece and not in bits like a chestnut.
C2: Oh, well it is then – press Y.
CBL + C1: Has the leaf got an edge like the teeth of a saw?
C3: My turn. Well, it's got a curved edge but not like a saw.
C2: Let's see – mmm, I agree.
C1: OK, I'll put no.
CBL + C1: Has the leaf an oval shape and the veins going right to the edge?
C2: Neither.
CBL + C1: Has the leaf got prickles?
C3: No.
CBL + C1: Oak.
C2: Let's tell T.
C3: It doesn't look like an oak to me!

Following the questioning attitude expressed by C3, T suggests further research in the library. Here the program acts as an initiator of other learning-related activities, and T is acting as a manager of the educational environment.

ACT 2: *In the library.*

C3: Oak leaves have deep rounded lobes with two small lobes at the base. Our leaf doesn't. The ends are more pointed.
C2: In an oak leaf the top part is wider than the bottom – ours is the other way around.

C1:	I think I've found it. Is it a maple leaf? It says that the leaf and veins have milky stuff in them, and this one does.

ACT 3: *Back with the Computer (The children teach the computer).*

The children confirm their finding with T and enter a new leaf in the SEEK data base. Still operating in the 'identification mode' which they were using in Act 1, they again get to the point where the program tells them that they have an oak leaf . . .

CBL + CI:	Oak.
C1:	No! Oh, what's happening? (*The screen display is changing at this point.*) It's disappearing. Wait, part of it's still on. It says 'What is it?'
T:	It's asking what the leaf is if it isn't an oak.
C1:	Oh, silly me. Right, let's put in maple. M-A-P-L-E.
CBL:	A question to give the difference?
T:	It wants you to tell it a question to show the difference between the two leaves.
C2:	Is the top wider than the bottom?
C3:	Does it have two small lobes at the base?
C1:	Does it have milky stuff?
T:	Which one do you want to use? Remember that it must have a YES/NO answer.
C3:	Use Jackie's *(C2)* – it's the clearest.
C1:	What do you mean?
C3:	Well, it's the easiest for anyone to see when they look at it. *(All three children are actively involved in devizing and evaluating questions.)*
C1:	OK, shall I type it in?
C2 + C3:	Yes.

The children then entered their newly-found defining characteristics and new datum, and worked through the program to check that they could identify a maple leaf.

Not all interactions with the computer will be of this quality but there is now enough evidence to show that the computer in the classroom can stimulate worthwhile educational outcomes, and this exchange illustrates a number of those desirable outcomes. A questioning attitude is apparent, with no straightforward acceptance of machine-produced solutions and a marked appearance of both child autonomy and group cohesion; the children appear to be teaching themselves, partly by teaching the computer; the teacher has a managerial role in that she no longer provides final answers, but rather provides pathways to answers for the children to explore themselves. In a chapter concerned with evaluation, perhaps the most encouraging development is the appearance of the children's questioning of their own questions – they come to understand what is a good question through evaluation.

Whereas some educational innovations seem to disappear very shortly after they are heralded, CBL will be with us for some time. It is too effective for us to ignore it. The problem discussed here concerns the nature of these effects – effects upon the social organization of the learning environment, and effects upon the growth of children's minds. Not only should we provide an account of these social and cognitive changes, but we should also provide an evaluation of the CBL styles which are socially and cognitively effective.

We have already commented on Amarel's (1984) disturbing observation that the less able are increasingly using drill-and-practice programs while their more able peer group are introduced to the computer as a tool. Just as this might result in a disparity in cognitive outcomes, we may also be faced with a compliant disadvantaged child whose more advantaged peer is encouraged to develop those social skills of leadership and active co-operation which are so prized by our society. These changes must be monitored.

In arguing the benefits of the computer in the cognitive life of the classroom, we are not saying that children learn more: we are arguing that with software which allows the computer to be used as a tool to extend and amplify children's thinking skills, they will be involved in a different education. This education is one that many would argue is more appropriate to a society demanding increasing flexibility from its members.

REFERENCES

Amarel, M. (1984) Classrooms and computers as instructional settings. *Theory into Practice, 22*, 260–266.

Barker, R.G. (1968) *Ecological Psychology*. Stanford: Stanford University Press.

Becker, H. (1982) *Microcomputers in the classroom: Dreams and realities. Report No. 319*. Baltimore, MD: Johns Hopkins University, Center for Social Organisation of Schools.

Bruce, B.C., Collins, A., Rubin, A.D. and Gentner, D. (1982) Three perspectives on writing. *Educational Psychologist, 17*, 131–145.

Bruner, J., Goodnow, J.J. and Austin, G.A. (1956) *A Study of Thinking*. New York: Wiley.
Bunderson, C.V. (1974) The design and production of learner-controlled courseware for the TICCIT system: A progress report. *International Journal of Man–Machine Studies, 11*, 5–24.
Chambers, J.A. and Sprecher, J.W. (1980) Computer-assisted instruction: Current trends and critical issues. *Communications of the Association for Computer Machinery, 23*, 332–342.
Chandler, D. (1984) *Young Learners and the Microcomputer*. Milton Keynes: Open University Press.
Char, C., Hawkins, J., Wootten, J., Sheingold, K. and Roberts, T. (1983) *The Voyage of the Mimi: Classroom Case Studies of Software, Video, and Print Materials*. Report to the US Department of Education. New York: Bank Street College of Education.
Chatterton, J.L. (1985) Evaluating CAL in the classroom. In I. Reid and J. Rushton (eds) *Teachers, Computers and the Classroom*. Manchester: Manchester University Press.
Clark, R.E. (1984) Learning from computers: Theoretical problems. Paper presented at AERA, New Orleans.
Cockcroft, W.H. (1982) *Mathematics Counts*. (The Cockcroft Report.) London: HMSO.
Daiute, C. (1985) *Writing and Computers*. Reading, Mass.: Addison-Wesley.
Hartley, R.J. (1984) Evaluation of a CAL package for modelling. *Computers in Education, 8*, 69–76.
Hawkins, J., Sheingold, K., Gearhart, M. and Berger, C. (1983) Microcomputers in schools: Impact on the social life of elementary classrooms. *Journal of Applied Developmental Psychology, 3*, 361–373.
Hoyles, C., Sutherland, R. and Evans. J. (1986) Using LOGO in the mathematics classroom. What are the implications of user devised goals? *Computers in Education, 10*, 61–72.
Johnson, D.C., Anderson, R.E., Hansen, T.P. and Klassen, D.L. (1981) Computer literacy and awareness. In J.A.M. Howe and P.M. Ross (eds) *Microcomputers in Secondary Education*. London: Kogan Page.
Kidd, M.E. and Holmes, G. (1984) CAL evaluation: A cautionary word. *Computers in Education, 8*, 77–84.
Lichtman, D. (1979) Survey of educators, attitudes towards computers. *Creative Computing, 5*, 48–50.
Nicolson, R.I., Bowen, P. and Nicolson, M.K. (1984) Classroom evaluation of 'SUMS' CAL arithmetic program. *Human Learning, 3*, 129–136.
Ogborn, J. (1985) Thinking with computers: The computer as a tool to do what the learner wants. Paper presented at the CAL 85 Conference, Nottingham.
Papert, S. (1981) *Mindstorms: Children, Computers and Powerful Ideas*. Brighton: Harvester Press.
Preece, J. and Jones, A. (1985) Training teachers to select educational computer software: Results of a formative evaluation of an Open University pack. *British Journal of Educational Technology, 16*, 9–20.
Ridgway, J., Benzie, D., Burkhardt, H., Coupland, J., Field, G., Fraser, R. and Phillips, R. (1984) Investigating CAL? *Computers in Education, 8*, 85–92.
Rubin, A. (1983) The computer confronts language arts: Cans and shoulds for

education. In A.C. Wilkinson (ed.) *Classroom Computers and Cognitive Science.* New York: Academic Press.

Sheingold, K., Hawkins, J. and Char, C. (1984) "I'm the thinkist, you're the typist": The interaction of technology and the social life of classrooms. *Journal of Social Issues, 40,* 49–61.

Smith, G.W. (1986) CAL: Improved learning and improved teaching. *Computers in Education, 10,* 115–118.

Suppes, P. (1966) The uses of computers in education. *Scientific American, 215,* 207–220.

Turkle, S. (1984) *The Second Self: Computers and the Human Spirit.* London: Granada.

Underwood, G. and Underwood J.D.M. (1986) Cognitive processes in reading and spelling. In A. Cashdan (ed.) *Literacy: Teaching and Learning Language Skills.* Oxford: Basil Blackwell.

Underwood, J.D.M. (1985) Cognitive demand and CAL. In I. Reid and J. Rushton (eds) *Teachers, Computers and the Classroom.* Manchester: Manchester University Press.

Underwood, J.D.M. (1986) The role of the computer in developing children's classificatory abilities. *Computers in Education, 10,* 175–180.

Wason, P. C. and Johnson-Laird, P.N. (1972) *The Psychology of Reasoning.* London: Batsford.

Part 3
SOCIETY AND THE NEW TECHNOLOGIES

The three chapters that constitute this section deal with a number of the broad social issues associated with IT.

Increasingly, it is being recognized that an understanding of attitudes towards and awareness of this technology is crucial if the technology is to be used effectively. In his chapter, Kemp considers the role of such factors, particularly in terms of the extent to which attitudes and beliefs can enhance or hinder the efficient implementation of IT. He argues that, if the best use is to be made of the new technologies, it is important to monitor evidence about the impact of the technology at the individual, group, organizational and societal levels of analysis.

Long considers the impact that IT will have on society from a specific viewpoint – from a consideration that, using concurrent technology such as telephones and satellites, IT can literally be brought into people's homes through their television set. He discusses the role of home-based transaction services such as Prestel, and the ways in which our current social patterns might be changed – shopping from home, for example, might alter patterns of shops available and the normal social interactions that take place within shops. More specifically, Long considers some of the cognitive problems that surround home-based shopping, and develops models to investigate these problems.

Finally, Frude extends some of the questions raised by Long concerning computing at home. He argues that home-based IT will have increasingly powerful psychological effects on individuals and on families. Many of these effects can already be predicted from understanding the impact of other technologies, such as video recorders, on patterns of social and family life. In this final chapter, Frude looks to the future and argues that much more needs to be understood about such questions before the final cost–benefit balance sheet of IT's effects on people can be drawn up.

ATTITUDES TO INFORMATION TECHNOLOGY

Nigel Kemp

The success or failure of information technology in our society is largely dependent on the attitudes of the people who design it, work with it, live with it and consume it (Mills, 1985). It is important to attempt to understand the ways in which information technology influences and is influenced by the thoughts, feelings and actions of individuals and groups in both work and non-work settings. It has been shown throughout this book that the impact of new technology is wide ranging, covering many aspects of our lives. This chapter considers the effects on the attitudes of people to information technology in general, and then the relationship between attitudes and a specific application of the technology. These attitudes are many and various, and they change with direct experience of the technology. First, however, the new technologies will be outlined briefly and also the ways in which attitudes are studied. The chapter concludes with a consideration of future developments of information technology, and the role that psychology can play in monitoring and influencing these changes.

Interestingly, 'new technology' is often thought of as an inclusive concept and it is assumed that the individual technologies share very similar characteristics. Essentially, there are considered to be two main forms of information technology. The first processes and distributes information, while the second major type uses the information to control other processes. However, there is an almost endless list of applications, ranging from new and adapted consumer products to uses in the commercial, industrial, military and service sectors. Thus, a global evaluation of information technology as 'good' or 'bad' is perhaps less useful than considering, in some depth, specific applications of the technology. Whatever type of technology is used, however, what is important from the point of view of this chapter is that people are involved in the design, manufacture, programming and use of them all. The new technology is not free of people, and the ways in which they react to it greatly influence its effectiveness.

THE NATURE OF ATTITUDES

The study of attitudes and the part they play in social and working life has long been of central concern to psychologists (Warr, 1978). There have been numerous attempts to define attitudes. As long ago as 1935 it was suggested that an attitude is 'a state of preparation or readiness for response' (Allport, 1935). This approach, which is still commonly adopted, explicitly assumes that the defining characteristic of an attitude is the disposition it represents for someone to behave in a particular way with regard to the object of the attitude.

Nonetheless, as early studies of the relationship of attitudes and behaviour indicated (La Piere, 1934), it is misleading to assume that a simple relationship will necessarily exist between attitudes and behaviour. Attempts to correlate general measures of attitudes with specific behavioural reactions have often been very disappointing. Brayfield and Crockett (1955), for example, concluded from a review of studies concerning the relationship between attitudes to work and job performances that no consistent relationship existed. Certainly, subsequent theorists have agreed that the behavioural implications of general attitudes are complicated (Schwab and Cummings, 1970; Thomas, 1971). Attitudes are probably best thought of as a result of people's capacity to evaluate: how a person makes judgements of right and wrong, good and bad, just and unjust can be integral to an understanding of attitudes to a particular issue. Seen in this way it is clear that attitudes influence the way we gather, interpret and evaluate information, and the way in which we react to a particular issue or topic. But while attitudes do influence (and are influenced by) behaviour, attitudes also reflect a person's beliefs and feelings. In considering how best to define the concept of 'attitude' it is important to recognize how, on the one hand, attitudes have a cognitive dimension, reflecting as they do beliefs that a person holds about the issue in question, and, on the other, an affective dimension, reflecting emotional reactions to the issue.

Attitudes are best thought of as hypothetical variables. They can only be inferred from what people say or do and are intended to reflect their values and beliefs and perhaps also the norms of the groups to which they belong. So, certain responses would be grounds for saying a person has a favourable attitude towards information technology, while others would be grounds for postulating unfavourable attitudes. Psychologists and others interested in the study of attitudes have expended considerable effort in designing appropriate psychometric techniques to measure them (see, for example, Cook *et al.*, 1981; Dawes, 1971; Oppenheim, 1982). The techniques used all depend on an inference process, typically, attempting to establish the extent to which people agree or disagree with a selection of attitudinal statements about a particular topic and classifying their attitudes accordingly.

A continuing focus of interest within the field of attitude theory has been the issue of attitude change. At one extreme, popular images of the power of brainwashing techniques suggest that people's attitude structures are somewhat malleable; at the other, images of 'resistance to change' suggest that people can be intransigent and stubborn in their outlooks. Attitude change theory needs to come to terms with a complex series of issues, requiring as it does an unravelling of the complexities of 'who says what, to whom, how and with what effect'. What seems clear is that attitudes are likely to be easier to change if they are based on poor information, are not associated with cherished social roles and do not reflect a person's key value structure. Attitudes acquired in people's formative years that reflect core values and are closely integrated within significant social roles are likely to be less flexible.

Research on attitude change (Jaspers, 1979; McGuire, 1969) suggests possible ways in which attitudes to information technology can be influenced. The communications model of attitude change (which emphasizes the characteristics of information source, message, channel and receiver), indicates that major variables open for manipulation include the information that is available in any particular situation and the message that is transmitted. Applied to the field of new technology this model emphasizes that information is needed not just on the technical features, but also on behavioural aspects of the technology. In short, the positive, clear-cut and revolutionary images of technical advancement need balancing, because, in reality, it is not like that. Information should be disseminated about how new technology actually is used, or can be used, and what the effects are on people and institutions (see the 'Technology Pack' produced by Sheffield City Council, 1984, for examples). In part, this is an objective of this book, that is, to raise behavioural issues which are important for the success or failure of technological change. A balanced perspective gives a more reasonable basis on which to form opinions, and from which to design socio-technical institutions which have a realistic chance of success.

In addition, a fundamental way in which people reassess and change (or bolster) their attitudes and beliefs is by an examination of their behaviour. We re-evaluate our beliefs on the basis of what we do. For information technology, this means that people should have access to and gain experience of computers. In a more general sense, if we are serious about moving into an information society, people must have access to resources and training facilities for them to develop the skills and knowledge necessary to cope with a highly technical society.

Furthermore, as well as technical training, there is a strong argument for involving people more directly in the decisions taken over how and why the technology is to be used, and over the social structure and institutions into which it is to be placed. Participation and consultation can be used to allay fears about change. Importantly, they also make for better change overall, as people are actively involved in decisions which crucially affect their lives (French and Bell, 1984; Mumford and Henshall, 1978).

In the remainder of this chapter attitudes towards information technology in general are reviewed, a specific area of application is examined, and the chapter concludes with a consideration of developments in the area.

ATTITUDES TO INFORMATION TECHNOLOGY IN GENERAL

Almost daily, it seems, we hear of new developments in information processing, sensing or handling. Frequently, information technology is promoted in the media as attractive, sophisticated and necessary for our continued economic development. Similarly, governments and business worldwide actively promote innovations in information technology in order to improve international and domestic competitiveness. The mood is such that whether or not we wish to use new technology, there no longer appears to be any option. To survive as a viable concern, industry and society must automate.

Government policy in Britain has been directed towards both developing and applying the new technologies, and in promoting acceptance by the general public and institutions. Large sums of money have been earmarked yearly for research and development activities; for example through the Alvey Committee of the Department of Trade and Industry. Moreover, specific programmes have been developed which focus on the 'office of the future' and manufacturing applications. However, it is generally recognized that the relative level of investment in technological innovation in Britain is considerably lower than that of our major competitors, for example the USA and Japan. Furthermore, it has been pointed out that British industry and commerce have been slow to take up the new technology and are conservative in their use, despite strong warnings that we should 'automate or liquidate'. Indeed, in order to raise the level of public and business awareness, 1982 was designated 'Information Technology Year' in Britain.

There are potential problems with such intensive and, in a sense, inevitable technological developments. Put simply, people may not find information technology products or jobs acceptable. In addition, they may be apprehensive about their ability to cope with computers, or fearful about the effects on employment levels. Furthermore, organizations and society as a whole may be unable or unwilling to adapt to the changes. Considerable concern has been expressed by commentators in this area about the possibility of massive job losses following adoption of the new technologies. As a consequence of this, some people foresee major social upheavals taking place which will occasion challenges to the present social order. In part, to counteract these fears, the emphasis at policy level and in public pronouncements has been on the potential benefits of the technology for individuals, organizations and society. Much less attention has been drawn to its potentially damaging features. Currently, the job displacement effects of information technology are minimal, although, given an increasing level of sophistication and integration of systems, it appears inevitable that there will be significant effects on employment (Northcott, Fogarty and Trevor, 1985).

In the following sections, some of the current attitudes to information technology in general will be examined through a consideration of the views of the general public, management and the trade unions, and unemployed people.

The general public

Several surveys have been undertaken during the 1980s by MORI (Market and Opinion Research International) and OR and C (Opinion Research and Communication) into the views of the general public towards new technology (Mills, 1985). In the first survey, in 1980, of a representative sample of 1,000 people in Great Britain, a majority of respondents agreed that major technological change was necessary and should be accepted and not resisted. However, at the same time these people were fearful that this change was to have dramatic negative effects on employment and skill levels. This double-sided attitude continued throughout the surveys of 1981 and 1984. In addition, there was some evidence that people directly affected by new technology at work reported that the changes had been for the better.

The most recent survey of public attitudes to new technology was conducted in June 1985 (MORI, 1985). A representative sample of 1,824 adults was interviewed in 134 constituencies in Great Britain. Again, the vast majority of people (84 per cent) saw technology as essential to Britain's prosperity, although as a nation we were seen as slow in adopting it. Over half of the respondents were prepared to accept cautiously the effects which information technology was having on their lives. However, a majority also wished to be consulted more on developments, and some saw technological innovation as a major threat to society; for example nuclear weapons and centralized information stores on citizens were opposed. Surprisingly, one in five of the public appear virtually untouched by new technologies. In terms of work, 42 per cent of people saw information technology as affecting their jobs, the majority reporting positive effects on skill levels, although pressure at work was said to have increased. Interestingly, most people were prepared to retrain to develop new skills, presumably as this was considered essential to obtain work.

Managers and trade unions

The overriding impression in this area is that there are few problems concerning the initial acceptance of new technology at the place of work. Although there are exceptions, particularly in the print industry, where major industrial relations issues have arisen, large numbers of employees in offices and factories have accepted the technology, as have trade unions and management (Mills, 1985; Northcott, Fogarty and Trevor, 1985). Senior management especially is committed to the use of technology to increase efficiency and productivity, and to give industry and commerce a competitive advantage in the market place. However, there is evidence of a low level of awareness of the opportunities available with information technology, and a lack of management expertise to exploit them. Moreover, concern has been expressed about possible job losses, particularly for 'knowledge workers' at middle and junior management levels (see Child, 1984; Piercy, 1984).

Trade unions and their leaders have generally not resisted the introduction of new technology, and this has been principally in order to protect the jobs of their members (Jenkins and Sherman, 1979; TUC, 1979). There is, though, very real concern over how the technology is used, or will be used following further innovations. To this end, new technology agreements have been written (see Northcott, Fogarty and Trevor, 1985; Williams and Moseley, 1985). These include arrangements covering consultative and negotiating machinery, job security, the terms and conditions of employment, job content and control and privacy issues.

The unemployed

A study of the computing needs of wageless people was conducted in Sheffield in 1984 (Smith, Fryer and Fitter, 1985). The study arose out of a perception that employed and unemployed people have different degrees of access to information technology. Given the current shifts towards an information technology society, this was seen as leading to an imbalance in the relative power of these groups to obtain employment. That is, the unemployed either did not have relevant skills or were denied the opportunities to develop them. Interviews with the unwaged people revealed a wide range of opinions on information technology. Basically, these reflected prosocial or antisocial images.

Prosocial beliefs were concerned with increased levels of personal confidence through the use of computers, an understanding of the potentially beneficial uses to which new technology could be put; for example in health and education services, and the possible release of people from repetitive, boring jobs. However, it was also noted that as well as changing jobs for the better, new technology could be used to deskill jobs and to promote job losses. Other 'antisocial' images portrayed information technology as being used to control and exploit the weaker members of society, to depersonalize the workplace, to be used for the invasion of personal privacy and, overall, as not living up to the far-reaching and benevolent claims made for it by the media, business and government.

Summary

I have shown that the majority of people hold mainly positive attitudes to information technology in general. This conflicts with the views of some commentators who expected much more resistance, particularly from the trade unions. The introduction of forms of information technology into work and society is largely unchallenged at the present time, and is accepted as inevitable and part of necessary change. However, beyond that, people hold views which are much more differentiated. There is concern about how the technology is to be used, who controls it and what its effects are likely to be. In short, attitudes become complex and may incorporate conflicting beliefs at this level of analysis (see also Breakwell, Fife-Schaw and Lee, 1985). In the next section I examine the nature of attitudes for a specific application of information technology.

ATTITUDES TO A SPECIFIC APPLICATION OF INFORMATION TECHNOLOGY

This part of the chapter examines the attitudes of employees in a large engineering company which is progressively introducing a flexible manufacturing system (FMS) into its assembly area. The eventual system will consist of automated assembly, storage and handling facilities integrated through computer control. The FMS will take several years to implement, thus the findings reported below represent the first part of a longitudinal study. The current focus of this study is on the cognitive and affective dimensions of attitudes, although the collection and analysis of performance and other behavioural data is underway. At present four different technologies are used in the assembly area (see Mueller *et al.*, 1986; Wall *et al.*, 1986 for further details).

All the jobs involve inserting components into printed circuit boards which are for subsequent installation into computers. The first type of technology involves manual insertion. The second technology is also based on manual insertion, but here groups of employees are organized on a flowline, where work is passed down the line for specific types of insertion. In comparison with the first type of manual job, this method of work involves a restricted range of insertions for each individual employee, and to some extent the work is paced by the flow of workpieces. The third technology is computer assisted. Here the operator does manual insertions, but a computer specifies what component to pick up and where to place it. The computer controls a carousel, which restricts what component the operator can select and shines a light to mark the correct location on the computer board. The fourth technology is computer controlled and is directly equivalent to a Computer Numerically Controlled (CNC) machine used for insertion tasks. Here, the machine itself inserts components into the boards working under the control of a computer programme which is downloaded from a host computer. Operators load and unload the machines, call up the programs, monitor and adjust the machine whilst it is working, undertake quality checks and rectifications and liaise with maintenance and supervision.

Operator attitudes

The interest here is in the attitudes held by the four groups of operators about new technology. It is important to note that those employees using computer-based technology had prior manual experience. However, the manual employees only had direct work experience of manual-based jobs, although they could observe and talk about the computer technology. Data were gathered through a survey of production employees (a 95 per cent response rate was obtained) and from interviews.

As would be expected from the previous discussion, a majority of employees were in agreement that automation was necessary for the firm to stay in business. Thus, most employees accepted the need for change, although they also agreed it meant job losses. Aside from this global evaluation, disagreements were revealed between those with experience of computer technology and those without. A majority of computer experienced operators believed that the new technology was more efficient and led to higher levels of product quality and consistency. Moreover, they described new technology work on the CNC machines as challenging and complex, rather than monotonous and boring, and requiring the use of a wide range of skills and many different work roles. Operators on the CNC machines were also enthusiastic about their work. Finally, new technology work was described as demanding but satisfying.

In contrast, manual operators saw new technology as producing boring work which was paced and controlled and resulted in poorer quality products. Interestingly, the worst jobs were described as those on the flowlines and the computer-assisted technology where employees were paced and, for the latter group, directly controlled by the technology. Stress levels, for example, were highest on the flowline. Both manual and the CNC jobs were seen as preferable and, overall, operators using the computer technology saw it as a 'good thing'.

This study suggests that attitudes towards new technology change with use of the technology. The more experienced the operator of an automated system, the more positive are attitudes towards it. Another important point to emerge concerned people's beliefs about the impact of technology *per se* and the relative influence of other factors at work. Employees firmly believed that managerial practices were the major determinants over the quality of their jobs and working life. Thus, social choices over skilling-up jobs versus deskilling, control versus autonomy, the delegation or otherwise of decision taking and enlarged or restricted work roles were seen as more important than the particular form of technology used.

This distinction between social choice and technological determinism has been discussed at great length (see Braverman, 1974; Clegg, 1984; Wood, 1982). One position discussed elsewhere in this volume is that the technology constrains social choices but it does not determine them. Thus, the choices managers make about how a particular form of technology is to be managed, supported and operated will influence to a large extent the nature of jobs that people do. The choices taken are not trivial, either for the individual or the organization and they involve both costs and benefits. A new technology job which gives employees responsibility and control over operational decisions, and which requires the use of a range of skills has been shown to result in high levels of motivation, job satisfaction and commitment to product quality. Conversely, a new technology job where the employee loads, unloads and minds a machine, and where control and responsibility are vested in the hands of specialists and managers has been associated with negative shopfloor attitudes and work behaviour (see Clegg, Kemp and Wall, 1984; Kemp and Clegg, in press). Of course, a balance has to be struck between these costs and benefits throughout the organizational system (Clegg and Kemp, 1986).

The findings outlined above mirror those obtained from previous investigations which have examined the impact of traditional technologies, for example the assembly line or group working, on employee attitudes and behaviour (Kemp *et al.*, 1983; Wall, 1986; Wall *et al.*, 1986). It is also clear from this literature that earlier proposals for the design of jobs which enhance the quality of working life for job holders are also relevant in the area of new technology (Hackman and Oldham, 1980; Kemp, Clegg and Wall, 1980). Finally, it is apparent that information technology cannot be evaluated in any meaningful way if it is abstracted from the social context in which it is located. We need to know the specific type of technology we are dealing with and the social system that surrounds it.

Manager's attitudes

Interviews with senior management in the factory representing both line and staff also showed a diversity of attitudes towards new technology. This contrasts with the often publically expressed opinion that management holds a unitary view about the benefits of new technology, and that planning and decisions taken over technology proceed on a rational basis (see Clegg and Kemp, 1986; Kemp, Clegg and Wall, 1984). Decisions taken about new technologies were seen to be political in nature and to reflect the vested interests of stakeholders over how the technology should be implemented and used. Managers disagreed over whether productivity and quality benefits would accrue, and whether the technology was 'flexible' enough to cope with production or product changes.

Managers expressed diametrically opposed views as to whether the technical system should be controlled by operators or specialists and the degree to which human intervention was feasible or desirable in the manufacturing process. For example, manufacturing and maintenance engineering, planning and control, production, and information technology project managers favoured specialist control. Quality control and engineering, personnel, training, technical support and stores were in favour of operator control (as was the major trade union). Not surprisingly, the lack of a common view has made integrated effort difficult to sustain. However, all managers do accept that change has to happen. They also believe that major benefits are to be gained from improved financial control (for example by a reduction in labour costs) and also through the development of a 'hi-tech' company to compete in the market-place. To combat the problems inherent in working with such a diversity of views, the company is currently formulating strategic plans for the development and integration of the human and organizational aspects of technical change.

Trade union attitudes

The study also examined the attitudes of the four trade unions in the organization. Findings showed that the unions were not opposed to information technology itself, but were specifically concerned with how it was to be used. The main issue for the largest union was who had control of the technology. Basically, new technology was welcomed so long as it was under the control of their operators and provided job security for union members. This position clearly competes with the wishes of the other trade unions representing specialist or supervisory staff. Jobs in these categories are more likely to be phased out in the company, given the increasing use of automated systems (for example, diagnostic equipment and automated storage facilities). Operator jobs, in comparison, will continue to exist (although the total number will be smaller) and if operational control is given to them it will mean that indirect jobs are expendable. At present, there is no open disagreement amongst the unions, although the issues are being discussed. A further issue for the largest union is to develop knowledge and experience of the technology as quickly as possible. This is so that the union can attain a powerful bargaining position for future technological developments as the full FMS is implemented. In order to have their position formally recognised the union is negotiating a new technology agreement.

Summary

Overall, data gathered during this study have shown that not only are there differences amongst managers, trade unions and shopfloor employees in their attitudes towards new technology, but also that a diversity of views exists within each group. Furthermore, major differences exist over who has or should control the technology, and how it should be operated, supported and managed. Attitudes are, of course, enmeshed in a social and political context. Finally, the evidence suggests that attitudes to new technology are more favourable for those people with experience of using it than for those who do not.

DEVELOPMENTS AND CONCLUSIONS

The argument presented throughout this chapter is straightforward. First, most people seem to hold positive attitudes to information technology in general. Second, these attitudes become differentiated, including both positive and negative beliefs, when specific applications of information technology are considered. Third, attitudes are more favourable given experience of the technology. Fourth, new technology should not be considered alone, that is abstracted from its social context. And finally, attitudes have behavioural implications. There are costs and benefits associated with the technology and how it is used. Negative perceptions of new technology may result in people being unable or unwilling to perform at their best and in the technology being used suboptimally. Overall, particular social choices may result in the quality of people's working and social lives being diminished.

A large part of this discussion of attitudes to information technology has been concerned with the world of work. This is unsurprising, given the technical investments in manufacturing and commercial organizations, and a developing interest in the behavioural aspects of new technology at work. Indeed, this emphasis is reflected in chapters throughout this book and in other publications concerned with, for example, manufacturing (Wall, Clegg and Kemp, in press), beliefs about robots (Argote, Goodman and Schkade, 1983), and the use of new technology in offices (Otway and Peltu, 1983). However, the impact of information technology ranges much wider than just industry and the office; for example, in health and education, in the home, in financial services and in retail and distribution, and it is important that the discussion is broadened to consider, briefly, the effects on those areas (Child *et al.*, 1984; Forester, 1982, 1985; Rowe, 1986).

First, however, it should be noted that there are problems involved in drawing conclusions and attempting to generalize from the analysis presented so far of people's attitudes to information technology. This relates to the degree to which information technology is used by people and institutions and its penetration into society. Most commentators agree that the use of new technology is not extensive at present and its effects not yet widespread. Thus, the 'Information Technology Revolution' (Forester, 1985) has hardly begun, its major effects are still to occur, and no clear consensus has yet been reached over what these effects will be (Rowe, 1986).

The discussion in previous sections was based therefore on relatively limited developments in information technology. For example, most organizations use the technology only in a stand-alone capacity. Thus, a word-processor is a replacement for a typewriter, and a CNC machine tool is a replacement for grinding or turning machines. However, a prime characteristic of information technology is that different forms can be linked together. Office systems can be networked and manufacturing systems integrated. Moreover, and of considerable importance, the technologies do not just have the capacity to be linked together within the spheres of industry, commerce, the services and home sectors, but can also be integrated across them. This process will fundamentally alter the nature of society. Moreover, this merging of previously disparate areas is presumed to be achievable within the next 25 years (Bessant *et al.*,

1985). Current forecasting literatures suggest that the implications of this 'information society' are both fascinating and frightening.

The effects on employment were mentioned earlier in this chapter. Basically, it was suggested that, as new technology becomes more sophisticated and integrated, then it is likely that there will be a dramatic increase in the number of people unemployed. This outcome is imminent unless there are corresponding structural changes; for example to working patterns and practices through job sharing, shorter working weeks, contract working etc., and through organizational restructuring or through the creation of opportunities in the service and leisure sectors. Major changes of this type are not yet underway. Currently, job losses are high and increasing in the manufacturing and office sectors. There is also evidence to suggest that women in clerical and administrative positions and the older worker are being displaced disproportionately by information technology. In this context, the detrimental effect of unemployment on an individual's well-being is well documented (Warr, 1984). However, these individual problems may become subsumed by the potentially catastrophic effects on society of mass unemployment.

The use and development of information technology in the service sector is a major step on the way to the information society. New technology is particularly appropriate here, because a large part of the work of these services is concerned with the processing of information. In the private sector, major advances have been made in banking and finance, insurance, retail and distribution, the hotel, travel and leisure industries, and in telecommunications and cable systems. For example, local and national networks are being set up to integrate information from electronic point of sale facilities in retail outlets, via distribution centres and thence to financial centres, where funds will be electronically transferred.

Similarly, massive data banks are being constantly introduced and updated in the public sector. Here the concern is with activities such as those found in the civil service, for example, internal revenue, passports and vehicle licences, and also in the education, health and welfare, police and community services. Finally, it is envisaged that the above services, particularly from the private sector, will become available and linked into people's homes. This will enable new communication opportunities, and facilities such as home shopping, banking, education and entertainment will become readily available. The vision here is of an information-rich society with an equality of access to facilities, and the voluntary participation of all people. In reality, the picture may be quite different.

Many important issues are raised by the development of such large, integrated systems. One particular issue concerns the invasion of personal privacy. Personal data which are maintained on common data banks can be accessed relatively easily and cross-referenced by private companies, representatives of the state and other interested parties. The provision of adequate safeguards in this area is immensely difficult. Similarly, the use of data banks which link together police files, social security and tax systems could lead to a closely controlled and monitored 'police state'. Here the picture is one of 'Big Brother' watching over us. Computer crime may also become a widespread problem. In the areas of health and the legal system problems may arise with regard to notions of responsibility and accountability if expert systems become commonplace. In addition, the use of information technology at home could disrupt existing patterns of family life. Home-working could lead to further inequalities between the sexes and to the disturbance of social interactions with peer and work groups; and the move to home shopping, banking, education and entertainment may promote social divisions if particular groups in society are unable to participate for financial reasons. Finally, the use of international data transfer, and the power of multinational companies and financial institutions may make a mockery of representative democratic freedoms if control over major national issues is located outside the country.

Of course, shots of optimism can be injected into the debate to counter these pessimistic predictions (Rowe, 1986). Information technology can help to eliminate dangerous and soul-destroying forms of work, to transmit information for educative purposes and to improve health and welfare provision. Indeed, *how* information technology is used is considered to be a matter of social choice. In order for us to make these choices, however, we need access to a wide range of information on the use and effects of new technology. As a contribution to this process it will be important for psychologists to monitor people's feelings about developments in the area of information technology; to investigate and attempt to understand the processes that are associated with changes in attitudes and behaviour; and, generally, to help in the design of a society which contributes in a positive way to the psychological welfare of its members.

REFERENCES

Allport, G.W. (1935) Attitudes. In C.M. Murchinson (ed.) *Handbook of Social Psychology*. Worcester, Mass.: Clark University Press.

Argote, L., Goodman, P.S. and Schkade, D. (1983) The human side of robotics: How workers react to a robot. *Sloan Management Review, Spring*, 31–41.

Bessant, J., Guy, K., Miles, I. and Rush, H. (1985) *IT Futures*. London: National Economic Development Office.

Breakwell, G.M., Fife-Schaw, C. and Lee, T. (1985) Survey of student attitudes to technology. *Science and Public Policy, 12*, 337–340.

Braverman, H. (1974) *Labor and Monopoly Capital: The Degradation of Work in the Twentieth Century*. New York: Monthly Review Press.

Brayfield, A.H. and Crockett, W.H. (1955) Employee attitudes and employee performance. *Psychological Bulletin, 52*, 396–424.

Child, J. (1984) *Organizations: A Guide to Problems and Practice*, 2nd ed. London: Harper and Row.

Child, J., Loveridge, R., Harvey, J. and Spencer, A. (1984) Microelectronics and the quality of employment in services. In P. Marstrand (ed.) *New Technology and the Future of Work and Skills*. London: Frances Pinter.

Clegg, C.W. (1984) The derivation of job designs. *Journal of Occupational Behaviour, 5*, 131–146.

Clegg, C.W. and Kemp, N.J. (1986) Information technology: Personnel where are you? *Personnel Review, 15*, 8–15.

Clegg, C.W., Kemp, N.J. and Wall, T.D. (1984) New technology: Choice, control and skills. In G.C. van der Veer, M.J. Tauber, T.R.G. Green and P. Gorney (eds) *Readings on Cognitive Ergonomics – Mind and Computers*. Berlin: Springer-Verlag.

Cook, J.D., Hepworth, S.J., Wall, T.D. and Warr, P.B. (1981) *The Experience of Work: A Compendium and Review of 249 Measures and Their Use*. London: Academic Press.

Dawes, R.M. (1971) *Fundamentals of Attitude Measurement*. New York: Wiley.

Forester, T. (ed.) (1982) *The Microelectronics Revolution*. Oxford: Basil Blackwell.

Forester, T. (ed.) (1985) *The Information Technology Revolution*. Oxford: Basil Blackwell.

French, W.E. and Bell, C.H. *Organization Development: Behavioural Science Interventions for Organization Improvement*, 3rd ed. Englewood Cliffs, N.J.: Prentice-Hall.

Hackman, J.R. and Oldham, G.R. (1980) *Work Redesign*. Reading, Mass.: Addison-Wesley.

Jaspers, J.M.F. (1979) Determinants of attitudes and attitude change. In H. Tajfel and C. Fraser (eds) *Introducing Social Psychology*. Harmondsworth: Penguin.

Jenkins, C. and Sherman, B. (1979) *The Collapse of Work*. London: Eyre Methuen.

Kemp, N.J. and Clegg, C.W. (in press) Information technology and job design: A case study on CNC machine tool working. *Behaviour and Information Technology* (in press).

Kemp, N.J., Clegg, C.W. and Wall, T.D. (1980) Job redesign: Content, process and outcomes. *Employee Relations, 2*, 5–14.

Kemp, N.J., Clegg, C.W. and Wall, T.D. (1984) Human aspects of computer-aided manufacturing. In *Proceedings of IEE International Conference on 'Computer-Aided Engineering', No. 243*. London: Institute of Electrical Engineers.

Kemp, N.J., Wall, T.D., Clegg, C.W. and Cordery, J.L. (1983) Autonomous work groups in a greenfield site: A comparative study. *Journal of Occupational Psychology, 56*, 271–288.

La Piere, R.T. (1934) Attitudes versus actions. *Social Forces, 13*, 230–237.

McGuire, W.J. (1969) The nature of attitudes and attitude change. In G. Lindzey and E. Aronson (eds) *The Handbook of Social Psychology, Vol. 3*, 2nd ed. Reading, Mass.: Addison-Wesley.

Mills, S.C. (1985) British attitudes to new technology. *Economic and Social Research Council Newsletter, 55*, 24–26.

MORI (1985) *Public Attitudes to New Technology*. London: Market and Opinion Research International Ltd.

Mueller, W.S., Clegg, C.W., Davies, R.D., Kemp, N.J. and Wall, T.D. (1986) Pluralist beliefs about new technology within a manufacturing organization. *New Technology, Work and Employment, 1*, pp. 127-139.

Mumford, E. and Henshall, D. (1978) *A Participative Approach to the Design of Computer Systems*. London: Associated Business Press.

Northcott, J., Fogarty, M. and Trevor, M. (1985) *Chips and Jobs: Acceptance of New Technology at Work*. London: Policy Studies Institute.

Oppenheim, B. (1982) An exercise in attitude measurement. In G.M. Breakwell, H. Foot and R. Gilmour (eds) *Social Psychology: A Practical Manual*. Leicester: The British Psychological Society and The Macmillan Press.

Otway, H.J. and Peltu, M. (1983) *New Office Technology: Human and Organizational Effects*. London: Frances Pinter.

Piercy, N. (ed.) (1984) *The Management Implications of New Information Technology*. Beckenham: Croom-Helm.

Rowe, C. (1986) *People and Chips*. London: Paradigm Publishing.

Schwab, D.P. and Cummings, L.L. (1970) Theories of performance and satisfaction: A review. *Industrial Relations, 10*, 408–430.

Sheffield City Council (1984) *New Technology: Whose Progress?* Sheffield: Sheffield City Employment Department.

Smith, L., Fryer, D. and Fitter, M. (1985) A study of computing needs of wageless people in Sheffield. Sheffield: MRC/ESRC Social and Applied Psychology Unit, Memo no 708.

Thomas, K. (ed.) (1971) *Attitudes and Behaviour: Selected Readings*. Harmondsworth: Penguin.

Trades Union Congress (1979) *Employment and Technology*. London: Trades Union Congress.

Wall, T.D. (1986) New technology and job design. In P.B. Warr (ed.) *Psychology at Work*, 3rd ed. Harmondsworth: Penguin.

Wall, T.D., Clegg, C.W., Davies, R.D., Kemp, N.J. and Mueller, W.J. (in press) Advanced manufacturing technology, and work simplification: an empirical study. *Journal or Occupational Behaviour* (in press).

Wall, T.D., Clegg, C.W. and Kemp, N.J. (eds) (in press) *The Human Side of Advanced Manufacturing Technology*. Chichester: Wiley.

Wall, T.D., Kemp, N.J., Jackson, P.R. and Clegg, C.W. (1986) An outcome evaluation of autonomous work groups: A long-term field experiment. *Academy of Management Journal, 29*, 280-304.

Warr, P.B. (1978) Attitudes, actions and motives. In P.B. Warr (ed.) *Psychology at Work*, 2nd ed. Harmondsworth: Penguin.

Warr, P.B. (1984) Job loss, unemployment and psychological well-being. In

V. Allen and E. van de Vliert (eds) *Role Transitions*. New York: Plenum Press.
Williams, R. and Moseley, R. (1985) Technology agreements in Great Britain: A survey 1977–83. *Industrial Relations Journal, 16*, 58–73.
Wood, S. (ed.) (1982) *The Degradation of Work*. London: Hutchinson.

INFORMATION TECHNOLOGY AND HOME-BASED SERVICES: IMPROVING THE USABILITY OF TELESHOPPING

John Long

Computers continue to be introduced widely into office and factory. A more recent trend has been the introduction of computers into the home. Wherever information technology is to be found, there is a need to ensure that it can be used. Making computers usable constitutes a major challenge for industry and for society. It also constitutes an opportunity for the behavioural sciences, and psychology in particular, to demonstrate how science can contribute to improving usability. The aim of this chapter is first to describe approaches to improving the usability of computerized systems, and then to illustrate the particular approach employed to make a computerized home-based teleshopping service easier to use.

The chapter begins with a description of computerized home services, such as teleshopping, and the difficulties experienced by people using the services for the first time. Three possible approaches to increasing the usability of computer systems are outlined. These are an applications approach, a theory approach, and a combined applications and theory approach. The combined approach is then described in detail. This approach attempts to ensure compatibility between the user's representations and those of the goods for sale, the task and the system. This 'representational' approach is then illustrated by research undertaken to help system designers improve the usability of computerized teleshopping services by identifying ways of increasing representational compatibility. Following a description of the approach, information technology in the home is discussed more generally. The range of domestic information technology is recognized as greater than teleshopping. However, it is argued that the representational approach is sufficiently general to accommodate the wider range.

HOME-BASED TRANSACTION SERVICES

Several forms of public information service involve the display of text and graphics on domestic television screens. These systems differ in their technology. For example, 'teletext' information is broadcast by means of television transmitters, and 'cabletext' information is delivered by means of switched cable networks. The emphasis in this chapter is on 'videotex', which usually operates via central computers linked to the domestic television receiver by means of a telephone line. Two-way communication with the central computers is made possible by an alphanumeric keyboard which the user operates to request information and to conduct transactions. On British Telecom's Prestel system, for example, transactional services include shopping, banking, ticket booking, message sending and software down-loading. Teleshopping will be the focus of the approaches to be described here.

At this time, the uptake of videotex services has been modest. However, there are obvious advantages to home-based services. For example, teleshopping could be more convenient than conventional shopping, since goods can be ordered at any time and are sent or transported by the supplier. A teleshopping service might be particularly useful for those without transport (Gateshead, SIS), the disabled, those uninterested in shopping (Guy, 1982), or those who cannot spare the time to shop. There may also be disadvantages – teleshopping leading to the further isolation of those for whom shops and shopping have a social as well as a supply function, the old for instance. There may be other consequences too, for example on the number and type of people employed in the retail industry and their conditions of work. Technological developments such as cabletext make it likely that facilities for teleshopping will become more available to the public, due to reduced cost (see Buckley, 1985). However, it is not clear whether these services will be usable by the general public, who may have little interest in learning how to operate them. Difficulties in using the service are known to exist and examples are presented below.

PROBLEMS OF USABILITY

In the course of the research to be reported, members of the general public were observed using Prestel's teleshopping service for the first time to make 'simulated' purchases of goods such as groceries, home-furnishings and consumer products. Sources of difficulty were identified which interfered with the completion of the shopping tasks. These difficulties can be associated with the goods, the task and the system. Each will be illustrated in turn.

The goods

The domain of the teleshopping service is the goods available for purchase, for example cutlery, sweaters, wine, etc. The system's description of the goods takes the form of text or graphics. Shoppers experience difficulty in using the service because either the description of the goods offered by the system is inadequate for their purposes or because the graphics are poor. The text descriptions are short, partly because the screen has a limited text capacity, for example: 'Roses red. Size 60 cms. Cost £10 a dozen'. For one user, this information failed to provide an adequate basis on which to make a purchase. As she pointed out '. . . they don't give you the kinds of roses, the names or whether they are floribunda, whether they have scent, whether they're double . . . .' The graphics on Prestel are rather crude and although they can be used to present pictures of the objects for sale, for example a vase, their quality is generally poor – in the words of one user more like a 'logo' than a 'picture'. The description may be potentially inaccurate, as illustrated by a user looking at an object which was supposed to represent a vase for sale: '. . . it couldn't be a vase because . . . there's no way to put anything in it. It's blocked off at the top . . .'. The real vase was not blocked off, but the graphical image appeared to be.

The shopping task

The second source of user difficulties was associated with the shopping task itself. For instance, to order an item, such as a vase, the purchaser is typically required to enter a number – termed the 'reference number' (for example 'R.412') into the system by means of the keyboard. Users either failed to understand the function of the reference number or they failed to remember or to note it when it was displayed (that is at the same time as the description of the goods). As a result, they did not have the number available when it was needed, that is at the time of ordering the item, and so had to search back for it through the goods' descriptions in order to find it.

The computer

The third source of user difficulties was associated with the operation of the computerized system. For example, to order an item, the user is required to fill in a type of order form which is called a 'response frame'. The response frame contains 'fields' for details of the item, such as price, reference number, etc. and details concerning the purchaser, for example, name, credit card number, etc. After providing the relevant information, users are required to press one of two keys, following this instruction, which appears at the bottom of the frame:

KEY 1 TO SEND, KEY 2 NOT TO SEND

Users misunderstand the instruction, thinking it refers to the dispatch of the item to them, when in fact it refers to the transmission of the response frame to the centralized computers. This misunderstanding can lead to unfortunate consequences. For example, one user remarked: 'If they don't send, it presumably means I have to go down to the record shop and pick it up.' Had she pressed KEY 2, the system would not have transmitted the order to the supplier (in effect cancelling the order-to-be), yet the purchaser may have inappropriately attempted to collect it (without it being there to be collected).

If home-based services are to be usable by the general public, difficulties of the kind described here must be eliminated or reduced. The services are unlikely to be successful unless the users are able to operate them adequately. In the following section, three psychological approaches to usability are described.

APPROACHES TO PROBLEMS OF USABILITY

Three psychological approaches to improving usability suggest themselves. The first is based on the shopping application itself, that is the system is made easier to use by learning about how people operate an electronic medium for the buying and selling of goods. The second approach is based on theory, that is established scientific principles are employed to improve usability. The third approach combines both of these by developing theory which is appropriate for making the buying and selling of goods easier. Each of these approaches will be considered in terms of a framework which makes it possible to identify the similarities and differences between them. The framework which will be used to compare the different approaches to usability was originally constructed to model 'ergonomic' activities and is described in detail elsewhere (see Long, in press). It is shown in Figure 1.

In the framework, the real world of *work* is comprised of *tasks*. It is contrasted with the world of knowledge, which is comprised of different sciences as well as non-science. Non-scientific knowledge covers all forms of experiential knowledge, both craft and personal. These two worlds are related by means of intermediary representations, and associated activities which transform (that is translate) them. Scientific knowledge is acquired from working situations using a number of transformations, two of which – *analysis* and *generalization* – are shown. Analysing a task produces an *acquisition representation* of the working situation, whose main function is to support a simulation of the real world for the purposes of generating new information. Generalizing the information from studies and experiments produces the models and explanatory principles of scientific knowledge. All transformations can be applied iteratively. Once acquired, scientific knowledge can be applied back into the work situation. It is applied by a number of transformations, two of which – *particularization* and *synthesis* – are shown. Particularizing scientific knowledge to a work situation produces an *applications representation*, for example in the form of design principles for systems or recommendations concerning their use. Synthesizing, that is applying these principles to tasks in the real world in line with criteria of usability, will make the tasks easier to perform.

Ergonomic activities involve all of the behavioural sciences including physiology, biomechanics, psychology, etc. This chapter, however, in line with the subject matter of this book is concerned only with the psychological aspects of home services. Illustration of other sciences' contributions to information technology applications have been described elsewhere (Long, in press). Further, the chapter is concerned only with approaches related entirely or partly to scientific knowledge.

An applications model

A model of the activities associated with the applications approach to improving usability described in terms of the framework appears in

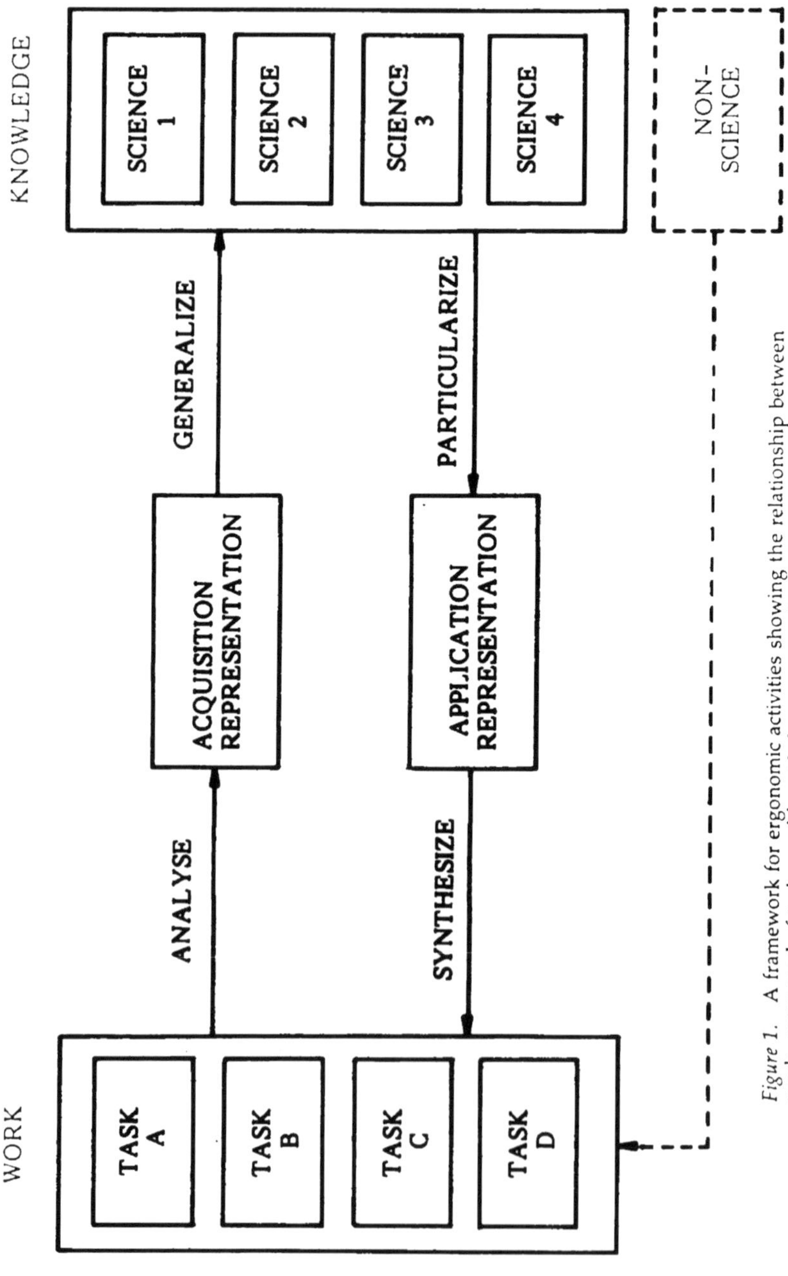

Figure 1. A framework for ergonomic activities showing the relationship between work, composed of tasks, and knowledge, composed of sciences.

Figure 2. In this case, knowledge is divided into science-based psychological knowledge and non-science based knowledge. Science is generally considered to be that type of knowledge whose phenomena (data), their measurement (methods) and explanation (theories/models) are required to be explicit and therefore public (that is 'objective'). In the applications approach, scientific psychology and non-science knowledge are likely to be inextricably mixed. Since the application is the driving force, a solution must be found, even if the relevant research has not been conducted. Further, the application of scientific psychology is implicit and informal, often based on intuition rather than formal theory.

However, a 'pure' applications approach to improving the usability of teleshopping services, pursued by a non-psychologist, for example a designer of videotex dialogues, would involve only non-science knowledge. The designer might try out the service for a range of goods and under a range of conditions (the acquisition representation) to identify problems in its use. The designer might then call upon experience to solve the problems. For example, to improve the quality of the descriptions of the goods, the designer might increase the amount of text used to describe roses (in the context of the difficulty described earlier) and to increase the number of colours in their graphical representation. These solutions – the applications representation – would then be used to direct modifications of the service itself.

In contrast, the applications approach pursued by a psychologist would be expected to include some aspects of scientific psychology as well as non-science knowledge. For example, psychological research has shown that some 'views' of objects are considered to be more typical than others, and as a result are more easily recognized (Rosch, 1975). A psychologist, then, might attempt to improve the quality of the teleshopping description of roses by offering different typical views, for example, of the whole rose and of the bloom.

The strength of the applications approach resides in its emphasis on solving the applications problem. The requirement to produce a solution is more important than for the solution to be based on scientific, psychological knowledge. Better ways of describing the roses offered for sale by teleshopping service could be devised even in the absence of Rosch's theory. This approach, then, is able to produce solutions even in the absence of relevant scientific, psychological research. Further, the approach is flexible with respect to time and to resources. Without using measuring equipment or even observing teleshoppers, the psychologist may be able to offer a solution

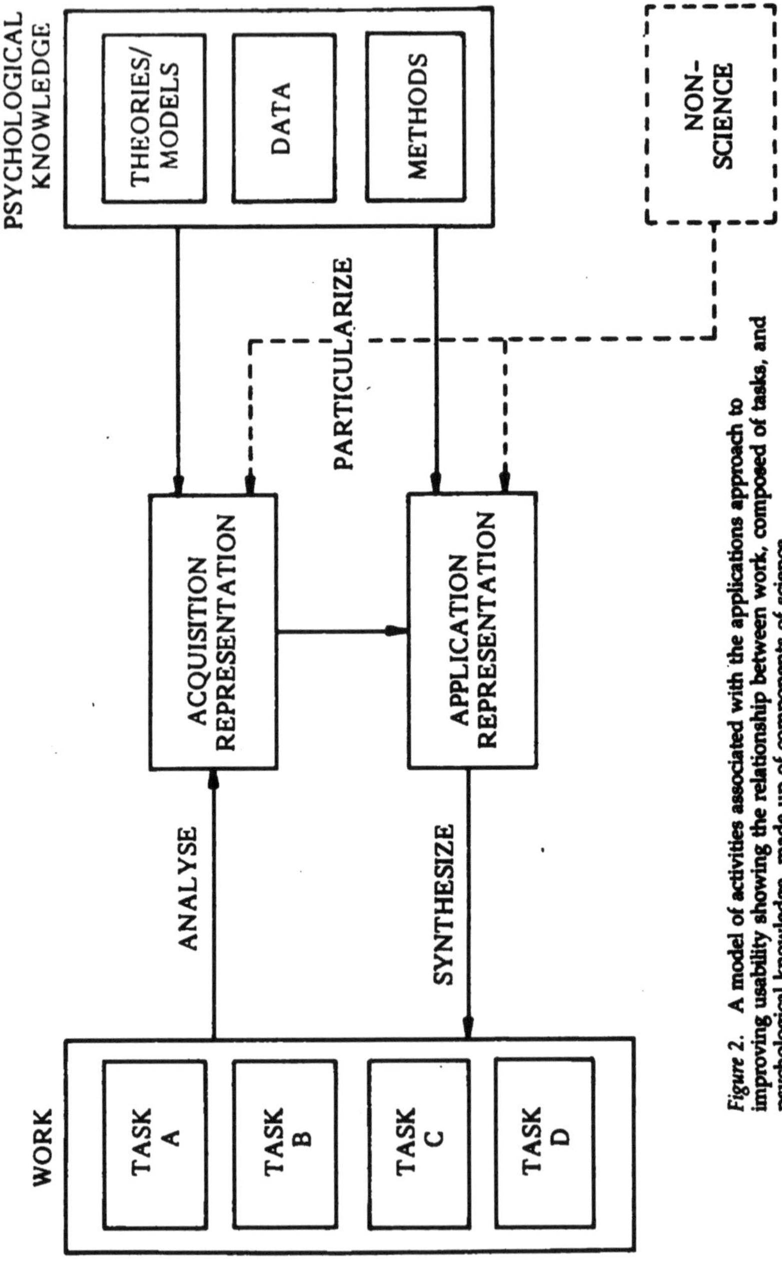

Figure 2. **A model of activities associated with the applications approach to improving usability showing the relationship between work, composed of tasks, and psychological knowledge, made up of components of science.**

to the problem of the inadequate description of the goods, based on experience and on using the system.

. However, the approach is not without its weaknesses. For example, the psychological content of the applications solution is likely to be limited and variable. At worst, there may be little or no psychological content for only inappropriate content. In the absence of Rosch's theory, the psychologist might have suggested simply increasing the amount of graphical detail to improve the description of the roses, much in the manner of the pure non-psychological applications approach, rather than providing alternative views of the goods. A further weakness is that of the efficacy of the solution. If the psychologist spends an insufficient amount of time trying out the system, if the service is directed at a new category of users, or if measuring equipment is not used to generate appropriate data, then even if a solution is forthcoming, its efficacy may be in doubt. For example, although the presentation of prototypical views of rose and bloom may be consistent with Rosch's theory, the solution may not be appropriate if the problem has not been correctly identified and if the system's graphics can only inadequately project prototypical views.

A theory approach

A model of the activities associated with the theory approach to improving usability is somewhat similar to that of the applications approach shown in Figure 2. The important difference between them, however, is that the theory approach involves only the application of scientific psychology. There is no application of non-science knowledge. In addition, the application is normally explicit. Psychology may be used to influence both the acquisition and the applications representation. Consider the second source of user difficulties cited earlier, involving the shopping task itself. Users experienced difficulty in ordering goods because they failed to note or to remember the reference number used by the system to identify the goods, and so were unable to fill in the response frame for ordering the goods. A psychologist pursuing a theory approach to this problem might consider it appropriate to apply Anderson's (1982, 1983) theory of skill acquisition termed ACT. This suggests that skills can be described in terms of simple 'conditional' rules such as:

> IF the goal is to order roses
> THEN note down the reference number of roses.

In the theory, skill acquisition has two stages. The first is a 'declarative' stage where a description of the skill ('obtain reference number') is interpreted by general rules ('note'), as described earlier. The second stage is 'procedural', in which the skill is directly embodied in the task-specific rules. For example, someone who orders roses frequently is likely to remember the reference number of the roses, and so does not need to note it down. The rule reflecting skilled behaviour might take the form:

IF the goal is to order roses
THEN enter R 642 in the response frame.

The reference number in this case is stored in the user's memory and can be recalled. Although the theory is more complex than these examples indicate, they adequately illustrate the theory approach for present purposes. On the basis of the theory, the psychologist might infer that new users' difficulties concerning the reference number might arise because they do not possess an appropriate rule at the declarative stage. They are able to note down details about the shopping service, but do not realize the need to do so with respect to the reference number. As a solution to the problem, the psychologist might suggest that the rule – expressed in a form suitable for users – precede the description of the goods. Alternatively, the psychologist might suggest that the reference number appear on the response frame, so obviating the need for the user either to note it down or to remember it.

The strength of the theory approach lies in its psychological content, to the extent that it is carried over into the applications solution. If the concept of a rule is well expressed in ACT, and is well supported by empirical evidence, and the theory is applicable to ordering goods via the teleshopping service, then a solution based on the theory should be successful in improving the usability of the service. Like the applications approach, the theory approach is also flexible with respect to time and resources. Its weaknesses, however, are threefold. First, if there is no psychological theory, then there can be no solution. Second, many theories are ill-formed or incomplete: it may in practice, then, be difficult to establish whether a theory applies to the tasks of interest. Third, if an inappropriate theory is applied to a problem, the solution may be inadequate.

A combined approach

The third approach to improving usability combines both of the above approaches and appears in simplified form in Figure 3. Like the applications approach, it is committed to finding a solution to problems of usability, whether or not appropriate scientific, psychological knowledge exists, although not by means of non-science knowledge. Like the theory approach, it exploits scientific psychology, but does not limit itself to current theories. In the absence of appropriate theory, data or method, the approach generates its own. Consider the user difficulty cited earlier involving the incorrect assumption that SEND referred to the goods rather than to the frame. A psychologist pursuing the approach might model the concept of 'topic' (of the dialogue between the user and the system), use the model to design alternative instructions to SEND and assess empirically the efficacy of the solution.

The strength of this approach lies in the psychological content of its models and data and their appropriateness for the particular problem. Its weakness is that it takes time, and requires equipment, etc. to measure and analyse the data. The approach is appropriate when the solution is applicable to many systems or teleshopping services.

Using the combined approach to improve the usability of teleshopping

The brief of the research which will be used to illustrate the combined approach, was to help system designers improve the usability of computerized transaction services. To this end, a research project was initiated to identify difficulties experienced by users and to suggest ways in which these difficulties might be overcome. A detailed model of the approach is shown in Figure 4.

The approach assumes that the difficulties experienced by users of computerized services result from a mismatch between the computer system's representations and those of the user (see Morton *et al.*, 1979). This is termed representational incompatibility. In the examples cited earlier, the representation of the goods in the form of the description of the roses, of the task with respect to the reference number of the goods and of the system concerning the SEND instructions can all be understood as incompatible with the representations of the user. Incompatibility may result in difficulties, because the user may be unable to relate the two representations appropriately.

The first stage in the research was to define its scope. The real world was sampled for current systems, tasks and users. These were characterized along dimensions likely to have consequences for usability (Gilligan and Long, 1984) and the scope of the research was

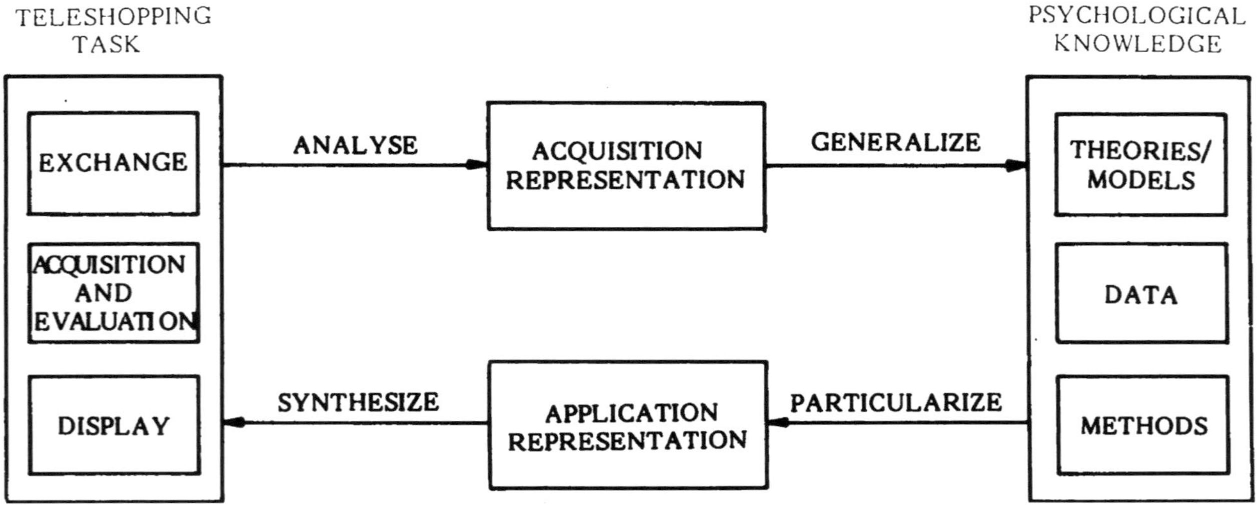

Figure 3. A simple model of activities associated with the combined approach to improving usability showing the relationship between the teleshopping task, made up of task components, and psychological knowledge, made up of components of science.

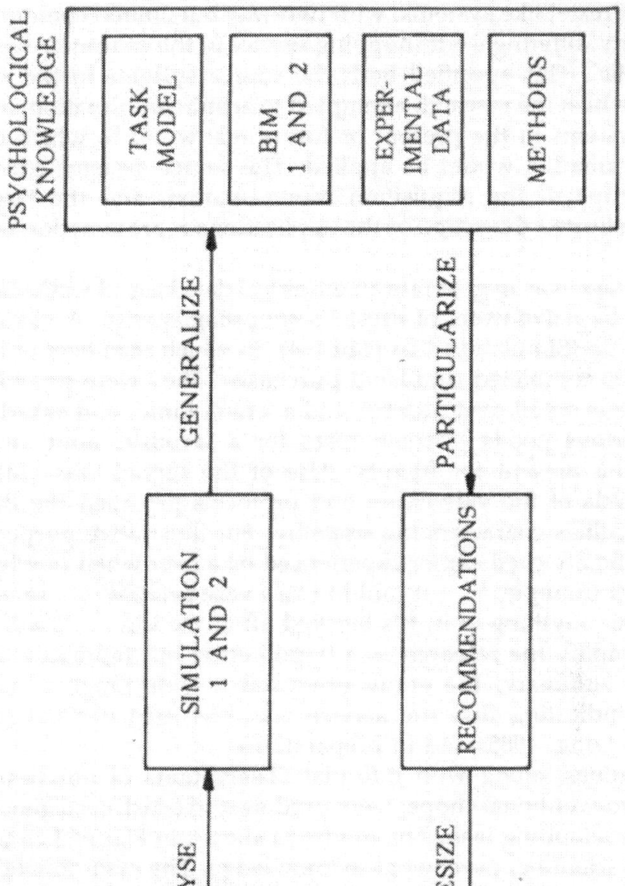

Figure 4. A detailed model of activities associated with the combined approach to improving usability showing the relationship between the teleshopping task, made up of source and target components, and psychological knowledge, made up of components of science.

defined as: 'Prestel-like systems, with two-way but limited communication capacity, offering a teleshopping service to the computer-naïve general public'. This specified both the source systems in the real world from which the research attempted to acquire information, and the target systems in the present or future real world to which the research information would be applied. The source systems constituted the scope of the acquisition representation, and the target systems constituted the scope of the applications representation (see Figure 4).

The next stage was to generate an empirical data base of difficulties experienced by naïve users of Prestel's shopping service. A simulation was set up (Simulation 1 in Figure 4), in which members of the general public completed simulated purchases. They were provided with a fictitious credit card number and a 'credit limit', and asked to purchase various goods – 'some roses for a favourite aunt' or 'a housewarming present for friends'. Use of the service was videotaped. Analysis of the videotapes and protocols provided the data base of difficulties characterizing usability. For illustrative purposes only, the difficulty cited earlier experienced by a user when purchasing a vase (for example, '. . . it couldn't be a vase because . . . there's no way to put anything in it. It's blocked off at the top . . .') will be used to exemplify the research as a hypothetical but representative example. In summary, the study produced a wide range of user difficulties, indicating that the service was less than optimal (see Buckley and Long, 1985*a* and in preparation).

The difficulties, along with informal observations of purchasing activities in conventional shops, were used next to construct a general model of the shopping task. The model as shown in Figure 4 constitutes part of scientific, psychological knowledge. The main functions of the model were: to organize the research in terms of specific task elements rather than only in terms of the overall task (see Figure 3), and to provide the psychological rationale necessary to relate changes in the system and in the user's knowledge of the service to performance of the shopping task (for further details, see Long and Buckley, 1984, and Buckley and Long, in preparation).

The fourth stage in the research was to generalize the individual difficulties identified by the study cited earlier, and to construct a model of the sources of user difficulties. The model is termed the Block Interaction Model (BIM) and appears in Figure 4 (BIM 1), along with the task model, as a generalization from Simulation 1. The model is an extension of the framework proposed by Morton *et al.*

(1979) and developed by Long *et al.* (1983). The function of the model was to identify those factors contributing to poor usability which might be modified to make the service easier to use. The model consists of the complete set of 'system' and 'user knowledge' variables necessary to categorize all the users' difficulties which were identified in the observational study. Variables were defined as factors which could be manipulated experimentally.

In the model, system and knowledge variables are associated with computer representations indicating 'ideal user' knowledge, that is knowledge required by the system for 'error-free' shopping. Difficulties in purchasing the vase can be understood as follows. Had the purchase been successfully achieved, the user's knowledge could have been considered compatible with the ideal knowledge required by the system's representation. When a difficulty is experienced, the user's representation is considered to be incompatible with the ideal user representation. Appropriate changes in the system and in the user's knowledge might be expected to increase the representational compatibility between them and the goods, the task and the system, and so to improve usability by removing the sources of user difficulties. For example, if included in the description of the goods were features of the vase used by the purchaser to evaluate the item ('cut-glass'; 'cheap'), the representation would be compatible. Likewise, if the user's knowledge included knowledge of brands ('Waterford'), even a poor graphical representation or a short text description would be adequate for evaluation.

The next stage was to assess by experimentation hypothesized effects on user shopping behaviour of changes in the system and knowledge variables identified in BIM 1. These effects must be demonstrated before the changes can be recommended to designers. The demonstration also constitutes a validation of BIM 1 by eliminating those variables which exert no significant influence on behaviour (hence BIM 2 in Figure 4).

A computerized simulation of a Prestel-like shopping service was developed based on the general task model (Simulation 2 in Figure 4). The simulation permitted manipulation of the variables identified in BIM 1. In one typical experiment, the system variable 'Extent of Description of Goods' was manipulated in terms of the presence or absence of criterial features for purchase in a text description (for the vase 'cut-glass'). The knowledge variable 'Knowledge of Transaction Domain' was also manipulated, by teaching users about features of particular goods ('the vase offered for sale by the system is cut-glass').

In the experiment, users were asked to evaluate goods according to a criterion which was specified. The time to complete the evaluation and errors committed in doing so were recorded by computer.

The effects of both experimental manipulations on evaluation performance were significant. For the system variable, including the criterial feature in the text description of the items to be purchased helped towards faster and more accurate evaluation of the goods. For the knowledge variable, teaching users about particular features of items to be purchased made the evaluation of those goods relative to those features faster and more accurate. These findings suggest that both the variables are likely to exert an influence on teleshopping behaviour. This justifies their inclusion in the validated BIM 2 (Figure 4) and in the recommendations for teleshopping service designers. Including criterial features in the goods descriptions and including goods whose features are known to users are both ways of making the service easier to use.

The final stage in the research was to communicate the findings to the designers of home-based teleshopping services (for details see Buckley and Long, 1986). The information took the form of design recommendations likely to increase the compatibility between the user's representation of the goods, the task and the system and those of the computer. The recommendations constitute the applications representation (see Figure 4). As an example, on the basis of the experimental findings reported earlier, it was recommended that to improve the ease of goods evaluation, criterial features – such as 'cut-glass' for the vase – should be included in the text. It remained for the system designers to apply the recommendations (the synthesis activity) to the target shopping services in the real world and for the users to experience the improved usability as a result.

This completes the description of the combined approach to improving usability. The approach is applications-based because its activity is motivated by a real applications problem – that of the user's difficulties in operating teleshopping services. It is theory-based, because the task model, the block interaction models, etc. are general to a class of system, task and user and not particular to any one.

INFORMATION TECHNOLOGY IN THE HOME

Although information technology has been more widely introduced into office and factory than into the home, domestic applications continue to increase. Prestel's home-based services, for example, also include: telebanking, by which money can be moved 'electronically' from account to account, and so be used to pay bills; telereservation of travel (rail, sea and air) and entertainment (cinema, theatre, etc.) tickets; telemessaging, by which electronic messages or 'mail' can be passed around a network of users; and the downline loading of software like videogames and educational packages. Many homes also currently possess their own stand-alone microcomputer which may offer a wide range of facilities for domestic use. If the home microcomputer is linked into a network, then services similar to those offered by Prestel can be accessed directly. Other domestic devices now involve microprocessors and include programmable central heating, washing machines, dish washers, etc.

New applications, however, can be expected to follow. For example, home-based microcomputers linked to networks will have access to 'electronic' libraries. Large informational data bases can also be accessed, for example the records of a local education authority. Purely 'electronic' organizations can be formed – perhaps for the housebound or elderly – by means of which help and communication can be made available. In addition, there is likely to be an increase in the use of computers to control home devices like central heating and electronic security and surveillance systems. The domestic robot – perhaps to identify and pretreat very dirty washing – would be a development of this type of application. Further, if 'expert' advisory systems develop, consultancy systems are likely to become available in the health field generally, and in other fields, for example law. Given these home-based applications, how generally appropriate is the combined approach likely to be?

There are a number of reasons why the combined approach applied to teleshopping will be generally appropriate for domestic information technology. The approach was able to deal with all the identified usability aspects of the domestic teleshopping service revealed by the research, including those associated with the task and the system and not just those associated with the goods as illustrated in this chapter. In addition, the approach has been applied to banking services (Wilson, 1984). Further, the approach – or modified versions of it – has been used to evaluate coding devices, including voice recognition for parcel sorting (Visick *et al.*, 1984); to develop a syllabus for training in information technology (Johnson *et al.*, 1984); and to improve the usability of computer-aided design systems for producing engineering layout designs (Whitefield, 1984). The approach has been demonstrated across a range of applications sufficiently wide to indicate its appropriateness for current and future domestic applications.

Further, the concept of a person as a processor of information, transforming representations of the real world into forms which are useful for behaviour, has been shown by psychological research to be a fruitful approach to characterizing and understanding human behaviour. The same concept underlies the present attempt to improve the usability of informational technology by identifying representational mismatches between the user and the system, by modifying the system or the user's knowledge with the aim of eliminating the mismatch and then by assessing empirically that the mismatch has indeed been reduced. The principles behind the approach, then, have very wide currency in and support from the discipline of psychology and as a result the approach itself is likely to be very generally applicable.

Lastly, the combined approach is by its nature sensitive to changes in systems, tasks and users. Unlike the theory approach, it is able to accommodate any application, even one for which there is no current theory. Unlike the applications approach, it is able to exploit theory or when necessary develop its own theory. In this way, the approach is both general and flexible.

In conclusion, then, there is every reason to believe that the approach applied here to improving the usability of a teleshopping service is more generally applicable to other present and future home-based information technology services and systems.

SUMMARY

This chapter describes a home-based teleshopping service and identifies three possible ways in which psychology can be used to make the service easier to use. One of the ways – a combined approach – is illustrated in detail. The illustration shows how difficulties in using the service can be reduced by presenting the goods in a manner compatible with the purchasing behaviour of the users. The approach attempts to improve usability more generally by increasing the representational compatibility between the system and the user. Other current and future home-based information technology services and systems are described. It is argued that the approach described is also likely to be suitable for improving the usability of these applications.

ACKNOWLEDGEMENTS

The research on teleshopping reported in this chapter was carried out jointly by Paul Buckley and the author. Paul Buckley also developed the task model. The research was funded by SERC and British Telecom under grant GR/C/23032. The opinions expressed are those of the author.

REFERENCES

Anderson, J.R. (1982) Acquisition of cognitive skill. *Psychological Review, 4*, 369–406.

Anderson, J.R. (1983) *The Architecture of Cognition*. Cambridge, Mass.: Harvard University Press.

Buckley, P. (1985) Realising the potential of viewdata. In N. Bevan and D. Murray (eds) *Man/Machine Integration*. Pergamon State of the Art Report 13/1. Maidenhead: Pergamon Infotech.

Buckley, P. and Long, J. (1985*a*) Identifying usability variables for teleshopping. In D. Oborne (ed.) *Contemporary Ergonomics 1985*. London: Taylor and Francis.

Buckley, P. and Long, J. (1985*b*) Effects of system and knowledge variables on a task component of 'Teleshopping'. In P. Johnson and S. Cook (eds) *People and Computers: Designing the Interface*. Cambridge: Cambridge University Press.

Buckley, P. and Long, J. (1986) Recommendations for optimising the design of videotex dialogues. In D. Oborne (ed.) *Contemporary Ergonomics 1986*. London: Taylor and Francis.

Buckley, P. and Long, J. Using videotex for shopping – a qualitative analysis. (Manuscript in preparation.)

Gateshead, SIS (n.d.) *The Gateshead Shopping and Information Service*. Gateshead, Tyne and Wear: Social Services Department.

Gilligan, P. and Long, J. (1984) Videotex technology: An overview with special reference to transaction processing as an interactive service. *Behaviour and Information Technology, 3*, 41–71.

Guy, C.M. (1982) *'Push Button Shopping' and Retail Management*. Papers in Planning Research 49, Department of Town Planning. Cardiff: UWIST.

Johnson, P., Diaper, D. and Long, J. (1984) Tasks, skill and knowledge: Task analysis for knowledge descriptions. In B. Shackel (ed.) *Interact '84*. Amsterdam: North Holland.

Long, J. (1986) A framework for cognitive ergonomics. In: D. Oborne (ed) *Contemporary Ergonomics 1986*. London: Taylor and Francis.

Long, J. (in press) Cognitive ergonomics and human-computer interaction. In P. Warr (ed.) *Psychology at Work*. Harmondsworth: Penguin.

Long, J. and Buckley, P. (1984) Transaction processing using videotex or: shopping on Prestel. In B. Shackel (ed.) *Interact '84*. Amsterdam· North Holland.

Long, J., Hammond, N., Barnard, P., Morton, J. and Clark, I. (1983) Introducing the interactive computer at work: The user's views. *Behaviour and Information Technology, 2,* 39–106.
Morton, J., Barnard, P., Hammond, N. and Long, J. (1979) Interacting with the computer: A framework. In E.J. Boutmy and A. Danthine (eds) *Teleinformatics '79.* Amsterdam: North Holland.
Rosch, E. (1975) Cognitive representations of semantic categories. *Journal of Experimental Psychology: General, 104,* 192–233.
Visick, D., Johnson, P. and Long, J. (1984) A comparative analysis of keyboards and voice recognition in a parcel sorting task. In E. Megaw (ed.) *Contemporary Ergonomics 1984.* London: Taylor and Francis.
Whitefield, A. (1984) A model of the engineering design process derived from Hearsay-II. In B. Shackel (ed.) *Interact '84.* Amsterdam: North Holland.
Wilson, F. (1984) The suitability and usability of videotex technology for supporting transaction processing. Unpublished MSc dissertation, University of London.

INFORMATION TECHNOLOGY IN THE HOME: PROMISES AS YET UNREALIZED

Neil Frude

In this chapter I am concerned to establish one main thesis – that current IT research promises to provide products which will have major psychological and social effects on 'ordinary users'. I am little concerned with the impact of currently available systems, because I believe that for 'ordinary users' this impact has been and remains relatively low. It must be expected that the effects of future systems will be of a completely different order.

The argument I put forward has the following form:

- ☐ Currently available IT systems have had relatively little direct impact on the 'ordinary user'.
- ☐ Powerful new systems are imminent.
- ☐ These will be swiftly implemented as consumer products.
- ☐ Such products are likely to have profound psychological effects.
- ☐ There is an unfortunate widespread reluctance to consider such effects in advance of the products themselves.

IT USERS TODAY

In one way or another everyone is a consumer of the products and effects of new technology. People watch television pictures which depend on special effects achieveable only by picture digitization. They read newpapers and magazines which may have been computer typeset, containing photographs which have been sent by facsimile transmission. They receive bills which are processed by mainframe computers. Their hi-fi system may be microprocessor controlled.

Most people, however, would not consciously identify themselves as users of information technology. Their consumption of IT is 'incidental' and they regard computers and microchips as part of a specialist and alien culture. If they appreciate the rapidity with which systems are being developed, then their mystification may be all the greater.

But knowledge about information technology, and access to systems, is not restricted to technologists and allied specialists. As well as those who are professionally involved in developing or implementing systems there are two other main user types. We can label these the 'workers' and the 'hobbyists'.

The 'workers' use one or more specialist systems regularly to achieve some work-related goal. They may operate systems on the factory floor or in a library, for example, or use a computer for word processing. The other group – the 'hobbyists' – use IT systems principally for recreational purposes. They may buy a computer in order to indulge in the 'hobby' of computing, or use it as a games machine.

One way of understanding why people differ in the degree to which they are attracted to IT systems is to analyse individual choice in terms of 'value'. People will opt to use IT systems as long as they get, or anticipate, value from those systems. The 'value' attributed to a system can be said to result from an implicit process of 'cost-benefit analysis'. If the benefits are judged as outweighing the costs then the system will be assigned a positive value.

This simple analysis can be used to suggest why 'experts', 'workers' and 'hobbyists' are attracted to IT systems and why others are generally not. 'Experts' would be expected to place a high positive value on IT because they gain numerous intellectual, professional and financial rewards from working with advanced systems. As a result of their experience and training they may also find it relatively easy to handle such systems, so that the 'effort costs' involved will be low.

Many 'workers' do not operate IT systems from personal choice but because the nature of their employment demands such involvement. Nevertheless, most of those who are offered an opportunity to work with the new technology are eager to take advantage of IT systems (Wainwright and Francis, 1984). They anticipate that use of such systems may bring financial and professional benefits, including higher status, bonuses and more satisfactory work conditions. IT may be seen as a useful means for achieving a worthwhile end.

For 'hobbyists,' interaction with an IT system can be an end in itself. They may create data bases for home accounts, records or books, for example, which could be handled more effectively with a simple card system. But they do not operate in order to maximize efficiency. For them, the processes of developing the system, and of inputting the data, are intrinsically rewarding. Any extrinsic pay-off, therefore, may be seen merely as a bonus. They may later use their skills to derive additional thrills. They may gain access to restricted databases – not to take advantage of the secrets contained therein but merely to experience the joy of 'cracking the system'.

Thus the expert, the worker and the 'hobbyist' are all prepared to invest some effort in interacting with IT systems. They are prepared to develop specialist skills and to learn something about the operation of the system. Such people, however, remain a minority. Most people are content to benefit from the application of IT as long as it does not require effort on their part. When special skills are required this brings a cost which, in the absence of apparent benefits, leads them to evaluate the system negatively.

One of the major 'costs' involved is that of learning to operate the system. For most people, the keyboard and the VDU are important alienating features of the new technology. The fact that a sequence of precise commands usually needs to be issued before any desired effect is obtained means that some degree of instruction learning is also required. The intellectual and keyboard skills needed to communicate with a system may be judged as difficult and laborious. And, in addition, many people have come to fear products which bear the hallmarks of 'advanced technology' and feel alienated by such gadgetry.

They are likely, also, to see IT systems as offering few potential benefits. What information available via, say, Prestel would be of use or of interest? To what practical use could they put a computer? How would it make their life easier or more enjoyable? Early advertisements for home computers boasted that the machine could be used to keep household accounts and 'run a computerized home'. But such claims were regarded with a scepticism which turned out to be fully justified.

Games playing and programming do often prove attractions, but these are frequently short-lived. It may be difficult to justify, even to oneself, the effort expended on programming when the results are so often patently trivial and sterile. Like completing a crossword puzzle, the process of programming may itself be challenging and absorbing, and the feeling of mastery rewarding. The resulting handiwork, however, is usually unsatisfying. Most people have few real 'jobs' for which the computer can provide help, and writing a suite of programs is in the majority of cases akin to making a set of gardening tools when you don't have a garden. At least you might find a market for the tools, but the products of home programming are rarely of 'merchandisable quality'.

In recent years the advent of cheaper and more powerful machines, together with sophisticated software packages, has renewed the promise that the home computer can be a useful tool. Once again, however, the problem is likely to be that users will quickly find that there are few real applications for their system. For the secretary and the professional writer, the word processor has become an indispensable aid, but most people will find that they actually have few words to juggle with their powerful word processing system. The number of home users who can really derive substantial benefits from this application is therefore limited. Following this line of argument, a pattern can be discerned. A clear majority of people are either ignorant of IT systems or judge that they have nothing to gain by coming to terms with them. Many of those who *are* initially interested have not been persuaded that domestic IT systems will really offer something useful. And those who have invested time and money in computers frequently become frustrated at the lack of useful applications. Countless thousands of parents must have purchased computers as 'educational tutors' for their children, later to be disappointed by the limited range of suitable software and the inflexibility of the machine as a learning system.

A number of other current IT systems which might have seemed potentially attractive to the ordinary user have also proved to have little impact. The reasons for this often stem from fundamental limitations in the technology, but not infrequently there have also been problems in how the technology has been implemented. A notable example of this is Prestel. For several years access to a very powerful data base system has been potentially available to everyone who has a telephone. The uptake of Prestel by 'ordinary users' however, has been minimal.

It is doubtful whether this can be explained by the financial costs involved or by the skills needed to access data. The problem has been, and remains, that the information accessible via Prestel does not have mass appeal, and that most of the information can be obtained more easily by other means. If Prestel provided information which was unique (for example, if it were developed as a magazine) or more easily accessed than alternatives (for example if an encyclopaedic knowledge data base were included) the system might attract users who were not finance professionals, travel agents or network 'hobbyists.' To prove attractive an IT system must offer something which is cost-effective compared with other potential sources, or which is unique.

Current IT systems have thus had a very limited uptake by 'ordinary consumers'. Although we all benefit indirectly from the implementation of systems by specialists (in medicine, publishing, design, etc.) few 'new' consumer products have proved attractive. Interactive systems remain 'inhospitable' largely because of the technical limitations which govern the nature of the person–machine interface. As a result, several product applications which have for some time been technically feasible have not fully developed commercially. At this point the costs for the ordinary user outweigh the benefits.

This seems likely to change. Future systems will certainly be (a) easier to operate, and (b) more 'useful'. If this is regarded as too 'optimistic' a picture of the IT future, support for such a view can easily be gained from a consideration of the 'agendas' of current technological research and of the indicators of commercial opportunism in the field. All of the evidence points to a future in which powerful and 'attractive' new systems will become technically feasible and will be developed commercially to be maximally appealing to the 'ordinary user'.

Because the IT future will certainly be radically different from the IT present, with respect to the systems available, their applications and the range of people who use them, information about the present use of systems, and their psychological effects, provides no reliable guide to the future. Before suggesting how the impact of IT is likely to change, however, we need to establish that radically different IT products will appear. To support this premise we need to consider the research agendas of those actively involved in the IT field and the commercial forces which will translate the fruits of such research into consumer products.

CURRENT RESEARCH AGENDAS

Research in the fields of computer technology and artificial intelligence is partly 'pushed' by an impetus to solve certain theoretical problems or overcome certain technical limitations. And it is partly 'pulled' by the potential of manufacturing desirable products. To appreciate the likely impact of current IT research it is necessary to distinguish between several different types of enterprise.

At one level there is a *technical* agenda for research – driven by a desire to expand basic technical parameters. Can information be processed more quickly, more efficiently and more 'intelligently'? Examples of work at this level include the development of new microprocessor architectures and the use of materials such as gallium arsenide to produce new generations of microchips. Fundamental research in software design can also be included.

At another level there is a *functional* agenda. How can machines be made to speak understandably, to listen intelligently, and to solve diagnostic problems effectively? Items on this agenda which are currently the subject of intense research include machine vision, natural language translation and applied knowledge engineering. Together, the technical and functional agendas constitute the major part of the large-scale IT research programes initiated in recent years in Japan, the United States, 'Europe' and the United Kingdom.

At another level there is a *product* agenda, much more the province of design engineers and commercial planners than of the technologists themselves. Implementing developments at the functional level, can an effective 'secretary machine' be produced to type directly from dictation? How can new developments be implemented in the form of more effective military weapons? Can we make a dish-washing robot? Can we produce an expert-system psychiatrist?

Finally, there is a research agenda regarding the resultant *effects* of technological innovation. Either in anticipation of a system, or following its introduction, questions will arise about its use and its impact. Many disciplines have interests in pursuing research at this level. Psychology is clearly prominent among these, but other relevant areas include the sociological, the strategic and the environmental.

There is a certain implied hierarchy in these levels of research. Developments at the functional level will largely depend on advances at the technical level, and product developments will clearly build on the success of functional research. This also suggests a temporal progression through technical, functional, product and 'effects' research. However, it is worth noting that to depict the relationships between these kinds of research as a simple linear causal path would be inappropriate and misleading. There are in fact numerous 'forward' and 'backward' links. For example, product ideas can stimulate research at the functional level. Also, and an issue of special importance for the current discussion, 'effects' research can anticipate developments at any of the other levels. Research at this level might well influence product development.

Prodigious advances have been made in recent years within the IT field, but it would be a mistake to be wide-eyed about these developments. This is because in terms of the currrent *agenda* for research, relatively little has yet been achieved. Compared with the advances which are anticipated, even by the most sober of technologists, we can say with some emphasis – 'You ain't seen nothin' yet'.

As IT systems develop they will become technically far more sophisticated. Paradoxically, their use will demand less sophistication of the user and they will thus become more 'approachable'. One way of conceptualizing this is to think of the user and the system sharing a common fund of interactive skills. As the systems themselves become more skilled, people will need less skill to interact with them successfully. Current research aims to make systems more sensible of the world and more intelligent; it therefore promises to break the current barrier between the non-specialist user and highly complex technical systems.

Besides being more 'approachable' at the interface, the powerful new systems which are envisaged will be able to offer the non-specialist user many more benefits than current systems. They will be more knowledgeable, more expert and far more flexible in their response style. Not only will they accept verbal commands and produce vocal output, but their knowledge base and their 'social' and 'teaching' skills will enable them to offer information in an attractive and entertaining way. Commercial exploitation will ensure that products emerge which will be able to present multi-media information in a style reminiscent of the tabloid newspaper and popular magazine as well as that of the financial broadsheet and the encyclopaedia. Thus the 'benefits' obtainable from the system will increase at the same time that the 'costs' involved in communicating with it decrease. Systems will no longer have value only for the 'specialists', but will appeal to the public at large.

Japanese technology planners have set out on a major project with the precise aim of producing ultra-powerful systems which are high on 'approachability'. In the original Fifth Generation proposal (ICOT, 1982) it was stated that the aim of the Fifth Generation project is to make the computer ' . . . a more amenable partner for people'. The Japanese have promised that operating such a machine would demand as little special knowledge of the user as a television or a washing machine. In their book 'The Fifth Generation' (1984), Edward Feigenbaum and Pamela McCorduck suggest that the Japanese:

aim to produce machines easy enough to use, and intelligent and

> fast enough in their responses, to come close to the kinds of transactions intelligent human beings are used to having with each other (p.17).

If the *process* of communication between the system and the person will become more 'natural', then there arises the issue of what will be the likely *content* of such communications. The question of what people would wish to communicate, and what information they would wish to receive, is psychological rather than technological. Commercial developers will need to know the answer to this question if they are to produce systems which are optimally attractive to non-specialist users. Even at this early stage there is compelling evidence to suggest that manufacturers will be eager to fully exploit the domestic product potential of conversational machines.

THE COMMERCIAL THRUST

Many developments which have an immediate and obvious application in specialist systems will also soon find application in domestic systems. A voice recognition facility developed by a company working on a defence related project, for example, may also be suitable for incorporation into children's toys. An oft-quoted example of such a 'spin-off' involves a microprocessor initially developed for use in an advanced weapon. The same 'dedicated' chip was eventually to be found operating as the control system for an alien attack arcade game. This example illustrates an important general point – that functions developed initially for business, industrial or military application will *incidentally* provide the means for new types of consumer products.

A major factor which will certainly lead to the early implementation of advanced systems in the domestic market is its sheer potential size. When compared to the market for personal computers, for example, the sales potential for non-specialist systems offering incomparably more benefits than present systems, for minimal effort cost, will certainly be enormous.

It is clear that commercial exploitation is very much in the minds of those who are coordinating the major national and international IT research programmes. Although many of the relevant statements on exploitation concentrate on specialist systems (in medicine, science, engineering, information science, etc.) there is a clear implication that any other applications would be welcomed. It is to be expected that entrepreneurs from many different fields will be encouraged to purchase the fruits of the technologists' labour and to incorporate them in any suitable product. Needless to say, this need not in any way detract from or inhibit the more 'serious' simultaneous specialist developments.

Indeed, the promise of a potential mass market for a particular development will encourage investment in areas which might otherwise be regarded as having too limited an application. An area in which such a symbiotic process has already been demonstrated is that of video cameras and camcorders. Although research and development aimed at the improvement of professional video cameras has done much to facilitate the emergence of effective home video systems, the lure of a potential mass market for equipment designed for a domestic video movie market has undoubtedly encouraged heavy investment in the area. And professional and specialist video systems have greatly benefitted as a result of this substantial investment.

The fact that the Japanese, in particular, have emphasized the universal 'accessibility' of their proposed Fifth Generation machine as one of its key characteristics implies an intention to place such systems in the hands of non-specialists. Indeed, many of their statements make this intention quite explicit. In this field, as in so many others, Japanese research and development never loses sight of the *product* agenda and the Japanese have consistently demonstrated their expertise in quickly incorporating developments at the leading edge of technological research into items for the mass consumer market.

Further evidence suggesting the rapid exploitation of IT developments comes from the intense involvement of manufacturers in the Japanese project. Companies which have invested money and personnel include Matsushita (who market products under the names 'National' and 'Panasonic'), Sharp and Fujitsu. These companies already produce robot systems for industrial application, but they are more familiar as manufacturers of domestic videos and hi-fi. It is unlikely that such companies are restricting their hopes for the application of Fifth Generation technology to industrial and business equipment.

Official statements regarding the product applications of Western programmes tend to be more restrained (or, it could be argued, less imaginative) than those of the Japanese, although some of the more colourful characters among the systems manufacturers have more than compensated for that. Thus in several recent interviews Sir Clive Sinclair has identified the domestic robot as a technically feasible product of immense potential, and the leading US robot manufacturer Joe Engelberger promised in a *New Scientist* interview: 'We'll have robots like ladies have hats'. Care should be taken not to lose sight, amongst such colourful rhetoric, of an important message implicit in such statements. Some manufacturers, at least, appear to be eagerly awaiting functional developments and foresee a huge mass market potential in their application. They read technical reports with more than an academic interest and observe laboratory bench demonstrations with an entrepreneurial eye. They want to ensure that they can make a sprint start in what they see as a major forthcoming race.

Others, however, have already jumped the gun. They have attempted to manufacture products before the necessary technical advances have been made. Not one but several 'domestic robot' models have been on sale for several years. 'Hero', 'Androbot' and their ilk may be primitive forerunners of a future range of useful and attractive products, but the adjective 'primitive' needs to be stressed. A domestic robot which cannot see, which 'hears' and 'speaks' only a limited range of set words and phrases, and which has nothing worth calling a 'knowledge base' is a pretty useless affair. Decidedly lacking conversation, these contraptions must remain, at best, conversation pieces.

Yet the introduction of such premature products clearly shows the restlessness of certain manufacturers and their faith that even highly imperfect 'embryonic' systems will capture a significant market. Recently, for example, Nolan Bushnell, the founder of Atari, introduced his 'Petsters' – a range of robot pets. The 'Petster De-Luxe' is a furry robot animal with infra-red sensors which responds to a limited range of vocal commands and repeats a number of simple sentences. The 'Petsters', however charming, are certainly premature products and at this stage of technological development they must stand beside other examples of product *concepts* awaiting the technological means for adequate product *realization*. And if such items are launched *in advance* of the facilitating technology, can it be doubted that important developments in the IT field will swiftly be implemented in the form of products designed for the mass market?

What kinds of products will these be? I believe that, even at this stage, it is possible to make reasonable judgements about some of the products which are likely to emerge as 'winners' in the future market-place. Such judgements need to take due consideration of likely technical developments, of course, but such information alone cannot predict product applications. We cannot therefore rely on technologists to be the best product forecasters.

WHAT TECHNOLOGISTS CANNOT TELL US

In designing, say, a new parallel architecture, the technologist is expanding technical parameters. The fruits of his or her efforts may be employed in an expert system, a data retrieval system or a real-time machine vision system. Thus the technical development may be used in a wide variety of *functional* applications. Furthermore, it may be incorporated into many different *products* – in a guided weapons system for example, or in a system designed to produce animated films. Technologists have little control over the application of their technical developments and may be poor predictors of products which will be realized as a result of their contribution.

IT can be regarded as a developing dynamic *medium*. Taking television as an imperfect analogy, we can see that, although technologists can predict for us some of the likely technical developments of that medium (flat-screen, stereo sound, etc.), they are *not* especially well placed to predict the future *messages* which the medium will convey. If we wished to predict whether the televisual diet of the future was likely to include more soap operas, less sport, etc. we would gain the best information not from technologists but from other sources. In particular we would turn to information about current consumer preference (the television ratings) and examine the intentions of the TV planners.

Following from this analysis we can construct a picture of the future of domestic IT-based systems in terms of 'medium' and 'message'. Technologists can advise us only on the likely technical and functional developments of the medium (or we can consult the relevant research agendas and assume that *some* progress will be made in at least *some* areas). Indeed, the major contribution of technologists in helping us to form a picture of the product future is their ability to inform us about the likely profile and time-scale of technological developments, and to caution us about what will *not* be possible. Within the limits of technical feasibility we will have to make judgements about the emergence of products according to other criteria. Commercial viability is clearly one of the most important of these criteria. Economic factors are of course important in determining such viability, but ultimately the survival and evolution of consumer systems must depend on what people will want.

WHAT WILL PEOPLE WANT?

Many products and services have the function of providing information and entertainment, and it is unlikely that the attraction of instant up-to-date information and of multi-media entertainment will fade. Future IT systems will surely provide products which fulfil such needs with more impact, more power and more personal choice than is now the case. The number of 'live' television channels will increase considerably, people will increasingly 'time-shift' with VCRs, and films, plays, documentaries and the like will be available 'off the shelf' or 'on-line'. Such developments do not await any technical advance. As systems become more 'intelligent', however, they will be far more flexible, with the ability to create new entertainments, for example, at the user's request and to the user's specifications. There would be no great problem in developing a system which came to predict an individual's preferences and which could therefore 'suggest' entertainments and proffer information which the user would find interesting and relevant. Systems would therefore permit greater initiative on the part of the user, but also show their own 'initiative'.

People now buy a wide variety of equipment to ease household chores. A vacuum cleaner is useful, but an intelligent vacuum cleaner would clearly be more useful. Dish-washing machines do not put away crockery and cutlery. One of the immediate applications aims of IT research is the production of industrial robots which are mobile and intelligent and have highly developed perceptual-motor skills. When such systems are first introduced they will of course be very expensive, but the recent history of product technology, in which phenomenal cost reduction has been achieved within a relatively short time-span, makes it likely that, when a suitably advanced generation of industrial machines is produced, their domestic cousins will not be long in gestation. The potential value of this market has been estimated as of the same order as that of the present market for automobiles.

But the intellectual and sensory skills which would be needed for a viable domestic robot would be so sophisticated and refined that to think of this product merely in terms of its dish-washing and dusting capabilities would be absurd. Compared with the costs of the macro-engineering components, supplying 'extra' intelligence and knowledge base would probabaly be cheap. Thus the robot designed for domestic use could be expected to possess a formidable range of intellectual skills and a widespread general knowledge (with updating capabilities) and to be an 'expert' in a variety of fields. The domestic robot would thus be able to play the roles of 'consultant', 'teacher', 'games opponent', etc. in addition to its more mundane role as domestic 'slave'.

One of the most fundamental human needs, and one which seems very far removed from the concept of 'product', is the need for social contact. People enjoy and thrive on the company of others. Usually these 'others' are physically present, but they may be on the end of a telephone line. And some of the contact which people experience as 'social' is not with other people but with pets. Thus not all 'social' contact takes the form of face-to-face contact with other people. It is not only possible but highly likely that advanced IT systems will find an important 'social' role by providing 'personal' conversational contact. I have elsewhere provided an account of the evidence which leads me to believe that this controversial development will in fact become a reality (Frude, 1983).

The vocal, conversational, responsive machine (particularly if it were also mobile) is not the kind of thing which people would easily regard merely as 'equipment'. They would certainly tend to anthropomorphize, and a body of evidence suggests (Frude, 1983) that they would enjoy treating the system as a social entity and would prefer such systems to be friendly 'creatures' rather than intelligent (but boring) tools. In a bid to produce as attractive a product as possible, therefore, manufacturers would include in the 'design characteristics' of such a machine, an attractive social style, and a 'character'. The skills of creative individuals – artists and playwrights working in a new medium – will be called upon to design 'friendship programs' and 'character software'.

Thus when the technical problems of speech synthesis have been solved and a system can speak 'naturally' (that is when the *medium* has evolved) it will of course be able to say *anything*. The accent, style of speech, vocabulary, etc. will be tailored to be especially pleasant. The freedom of the vocal 'palette' will be unrestrained, allowing designers to cause the machines, when appropriate, to whisper, laugh and maybe even cry. Endearments, witticisms and perhaps even mild obscenities will be uttered with the appropriate inflection. Having supplied the necessary medium, the technologists will not be called upon to supply the 'script'.

Such apparently trifling matters as the 'emotional authenticity' of verbal intonation will in fact be assigned great importance by the successful manufacturers, for they will appreciate the fundamental importance of the 'charm' factor in selling this new kind of artefact. Since the various brands of machine may all rely on the same fundamental components and have the same range of basic functions (as many calculators do today) such 'marginal' features (as well as the cost) may be vital in determining which models sell most successfully.

One of the necessary features of an ideal 'social' machine would be its ability to learn from experience. Not only would it be able to learn directly from the environment (including via the media) by watching, acting and drawing inferences, but it would also ask questions of the user in order to form a more complete picture of events, attitudes, beliefs and interests. It would be able to appreciate signs of approval and disapproval and would be sensitive to subtle nuances of tone and phrasing. It would learn to 'fit in' with the family and would acquire the 'manners' which the user found most acceptable. In a process of 'socialization' such as occurs with a child, the robot would 'mature' into the situation. Its 'mature personality' would be a product of both a 'nature' component – the fundamental characteristics of the model as determined by the designer – and a 'nurture' component – the result of the individual machine's experience, education and socialization.

THE EFFECTS

Products such as those envisaged here would certainly have important psychological and social effects. And such developments are likely to take place against a rapidly changing social scene. Largely through the influence of technological change there will be widespread movements in the pattern of employment, with fewer jobs and more people working from home. Technology will also bring about changes in the practice of medicine, law, politics and commerce. Demographic changes, especially those affecting family life, may be expected to continue.

Assessing the effect of one source of change against a dynamic social background is particularly difficult, but there is a need to consider the likely effects of the introduction of the kinds of products outlined. Some of the psychological questions concern 'immediate' effects – how, for example, would people respond to health education delivered by a personal robot? – while others concern more general factors – how will the new media change people's cognitive styles (Greenfield, 1984)? Two areas of special importance seem to be the effects of decreasing 'work' and increasing 'leisure' in a world with phenomenally enhanced information accessibility, and the social effects of systems with 'artificial personality'.

One utopian vision which might be derived from the prospect of major advances in IT systems is that of a population increasingly devoted to the pursuit of leisure and knowledge, freed of the tasks of labour and plugged in to a kind of multi-media juke box which could deliver, in surround sound and holographic imagery, any music or film from a universal library, any published literary work or scientific treatise presented in any one of a wide variety of forms and styles. Newspapers from around the world, instantly translated and read aloud, could be brought daily into the home. Encyclopaedic knowledge bases with superb teaching skills would be able to present personally tailored tutorials on any subject, at almost any level. Such a vision relies on certain premisses which are political and psychological rather than technological, and certain of these premisses are open to challenge. Would the economic arrangements in a future society provide most members with the means to access such facilities? And would people derive long-term satisfaction from such pursuits?

Systems with 'artificial personality' could be expected to exert a powerful influence on their users. There is a basic psychological tendency to treat certain kinds of objects anthropomorphically. The phenomenon of 'animism' has been elaborated with special reference to existing computer systems (Frude, 1983; Turkle, 1984) and it is clear that many of the ways in which artificial systems will in future be enhanced will render them highly susceptible to such a response.

The attainment of intelligent vision by a machine, for example, is likely to affect profoundly the way in which we human beings view that machine. A robot which 'sees' objects and reacts accordingly will have escaped one of the constraints at present embedded in our notion of a machine. It will behave as if it were 'aware' of its environment, and such 'awareness' is normally ascribed only to living objects. Evidence of visual alertness is one of the criteria by which we judge that something is conscious and alive.

There is a good evidence that people enjoy responding to artificial systems anthropomorphically and are attracted to systems which encourage such a response. Features which facilitate animism, therefore are likely to be deliberately promoted by product designers. Once a system is able to speak with natural intonation, for example, the designer (here a kind of 'script writer') will be at pains to exploit the potential 'human-ness' and 'charm' of the system. Some versions, at least, would also be produced with an outer form designed to enhance such a response. Initially we might expect animal-like forms, but consumer preference (as well ergonomic considerations for the practical machine) might quickly lead to an evolution towards the humanoid. The appearance of such a creation, as well as the tone of voice, might immediately suggest an age, a sex and certain stereotyped aspects of character.

A number of social and psychological effects might be predicted as a result of the introduction of such products, including a reduction in the problems of loneliness (pet animals have been shown to have an important effect on this), and an increased 'social' dependency on the machine. In my book *The Robot Heritage* (Frude, 1984) I discussed some of these issues at length. I also argued that a useful initial data base for deriving hypotheses about human–robot interaction is the extensive body of science fiction writing which addresses the theme. Thus the possibility that people might be at first horrified by 'social' machines and that manufacturers might have to design them with special features to make them initially acceptable, has been dealt with thoroughly in that literature. Similarly, much attention has been paid to the possibility of people becoming 'over-involved' with an intimate machine.

TRIVIAL PURSUIT?

There is currently an international race to produce 'intelligent machines'. The race is being backed in a number of countries by government investment, there are large-scale integrated programmes of research and many alternative paths are being explored simultaneously in a bid to achieve each of the various goals. The amount of intellectual effort being expended in this enterprise is unprecedented, and some of the best minds from a number of quite different disciplines are engaged in one way or another on what is effectively a global project. Clearly there is a widespread belief among the well-informed that the various goals on the technical and functional agendas are achievable, and in the forseeable future. It would be preposterous to assume that little will come of all this. Let's say that there would seem to be a 'better than evens chance' that the current efforts will result in systems which effectively reach the desired criteria in at least some of the key 'function' areas.

As new levels of competence are reached in the various functions, new systems will be made possible. We would not need to wait for the 'perfect' natural language translator for example before developments in that area could be usefully employed. Indeed, in all the functions being researched (voice recognition, machine vision, etc.) there are already systems which work at least at a low level of competency. In some areas we may expect steady progress, aided by developments of the basic 'technical parameter' level. In other areas there is a clear need for conceptual and technical 'breakthroughs'. But it is to be expected that any functional development which amounts to a substantial gain will be implemented. There are product concepts in industry, science, defence and business which are awaiting such developments.

In this chapter I have attempted to show that the achievement of IT research goals will also 'incidentally' result in powerful domestic products which will capture a wide market and have profound psychological and social effects. Consideration of the relevant technical, commercial and consumer issues led to the conclusion that the domestic robot is likely to emerge, since it represents in many ways the 'ideal' product realization of the agendas of high technology research.

Such a prediction might well cause dismay and disbelief, not least among some technologists. Caution is their watchword and, still working on the problems of realizing the fundamental technology, they may easily become irritated with those seen to take the success of difficult projects for granted. They are likely to dismiss such prognostications as 'presumptuous' in both of the key meanings of that term. Doubtless when polymer scientists were studying the molecular chains which produced new synthetic fibres they would have been irritated by those who 'assumed' that this might change the style of ladies' hosiery. Gutenberg, struggling to produce his Bible, would not have been amused if a fifteenth century psychologist had suggested that his technology might lead to tabloid newspapers and *Playboy*.

The feeling seems to remain that it is somehow discourteous to speculate about the potential product applications of a yet unrealized technology. But there is noticeably far less reticence about predicting the future weapons systems, for example, than of children's toys. One application is 'serious' and the other 'trivial'. Discussion about 'content' rather than 'process' (or 'message' rather than 'medium') is especially proscribed, with commentators on such aspects frequently dismissed as sensation-mongers (the current term of abuse for those who dare to conjecture in the IT field is 'wow merchants').

Technologists are not alone in their antipathy towards such projection. In a recent edition of the *Journal of Social Issues* devoted entirely to the impact of advanced technology on social patterns and behaviour, a social scientist (Caporeal, 1984) wrote:

> In those heady times it was predicted that computers would play grand master chess and discover new mathematical theorems.

> They would translate foreign languages and solve complex problems . . . The idea of 'thinking machines' had no shortage of critics arguing about what machines could not do (Dreyfus, 1979), should not do (Weizenbaum, 1976) or would do (Mowshowitz, 1976). Yet, for the most part, the controversies, even in the academic literature, were based on conclusions and speculations that far outstripped the data (p.17).

Curiously, the catalogue of goals ridiculed by the writer actually reads like extracts from the current functional research agenda for the programmes being funded by government bodies in Japan, the US and the UK. What one person may describe as 'speculations which far outstrip the data' another might describe as 'discussions about probable future developments'. The pace of innovation and application is now so rapid that it would be reckless *not* to anticipate, and irresponsible to merely 'wait and see'.

The IT future is predominantly a *product* future, and a major market for IT products will be the domestic market. They will evolve quickly, the survivors being those best fitting the needs and preferences of the consumers. In the haste to produce systems with 'artificial intelligence' for specialist applications we may brush aside too easily the notion that the consumer 'by-products' of such developments will also have 'artificial personality'. Those responsible for shaping such 'peripheral' aspects of product systems will not be technologists but designers, and their task will be easy when compared to the profound problems confronted by technologists today. But we should not make the mistake of equating ease with insignificance. And we should not make the error of assuming that an attempt to consider product issues is merely a trivial pursuit.

REFERENCES

Caporeal, L.R. (1984) Computers, prophecy, and experience: a historical perspective. *Journal of Social Issues*, *40*, 15–29.

Dreyfus, H.L. (1979) *What Computers Can't Do*. New York: Harper and Row.

Feigenbaum, E.A. and McCorduck, P. (1984) *The Fifth Generation*. London: Michael Joseph.

Frude, N. (1983) *The Intimate Machine*. London: Century.

Frude, N. (1984) *The Robot Heritage*. London: Century.

Greenfield, P. (1984) *Mind and Media: The Effects of Television, Computers and Video Games*. London: Fontana.

ICOT (1982) *Outline of Research and Development Plans for Fifth Generation Computer System*. Tokyo: Institute for New Generation Computer Technology.

Mowshowitz, A. (1976) *Inside Information: Computers in Fiction*. Reading, Mass.: Addison-Wesley.
Turkle, S. (1984) *The Second Self*. London: Souvenir Press.
Wainwright, J. and Francis, A. (1984) *Office Automation, Organization and the Nature of Work*. Aldershot: Gower Press.
Weizenbaum, J. (1976) *Computer Power and Human Reason*. New York: W.H. Freeman.

POSTSCRIPT
Information Technology and People: A Fresh View of the Relationship
Frank Blackler and David J. Oborne

A NEW APPROACH TO INFORMATION TECHNOLOGY

This book demonstrates the importance of considering people as significant components within the design, manufacture and application of information technology. As such it argues for a growing need to understand what people require from IT, how they can use it and how they do use it, and what effects it can have on them. This postcript enlarges on these points in turn.

Superordinate to such a thesis, however, is the philosophy that surrounds the concept of the person at the centre.

- How should people be viewed in the context of the new technologies?
- What will be the most useful approach to understanding their wishes, capabilities and requirements?
- How can the technology be harnessed to increase the user's efficiency?

As becomes apparent later, central to this must be an understanding of people's reactions to IT.

History has taught us that public awareness of an economically significant new technology is likely to pass through various, quite well-defined stages. Initially, there is likely to be a general indifference, as the technology is only vaguely reported, it appears remote from the needs of most people and its potential is poorly understood. Only later, as the advantages of the development begin to become more apparent, may people be galvanized into action. Then a period of frenetic activity can develop as efforts are made to adopt the technology quickly and to make up for any lost ground. At this stage both the benefits of using the technology and the disadvantages of not using it are likely to be overstated. However, as people begin to learn more about the technology a more sober and purposeful orientation will appear. Characteristic of this stage is debate about alternative approaches that may be adopted to the development of the technology and attempts by those affected by applications of it to ensure that it will be used in their interests. Only at the outcome of this key phase will the technology finally become integrated and accepted within a modified social and organizational order.

The same course of events can be seen in the IT field. Within the UK there was a general apathy towards microelectronic technologies in the early and middle 1970s. Towards the end of that decade public awareness of the technology began to develop. Given the importance of technological advance to the country's success in generating wealth, economic growth and international competition, policies were vigorously developed to alert people to the economic potential of microelectronics. Slogans such as 'modernize or fossilize' and 'automate or liquidate' became catch-phrases in certain quarters. Employers were urged to overcome the problems of adjustment that were anticipated, and suppliers to develop innovative products to meet the anticipated demand.

Now, in the late 1980s, a more sober and informed orientation towards information technology is beginning to appear. Managers have become aware that the mere purchase and commission of IT is, in itself, no guarantee of success. A slump has recently occurred in the demand for certain information technology products. Articulate professional groups who are beginning to be affected by the technology are influencing the ways in which it is coming to be used. Trade unions, too, have shown a developing understanding of how their members' interests can be damaged by certain types of applications of the technology. Civil liberties groups have pressed for safeguards in certain applications. People are now beginning to realize that information technologies are no easy fix for organizational, social and economic well-being. Suppliers are being forced to recognize that it is no longer inevitable that IT products will sell simply because they are packaged as 'state of the art' technology.

The debate that is now emerging about how IT should be used revolves around four interrelated issues. These are:

1. Which aspects of microelectronics-based technologies should be developed?
2. What goals, and whose goals, should they be harnessed to serve?
3. How should the processes of technological development, product design and systems application be organized?

4. How important are behavioural outcomes of IT, and what behavioural and social assumptions should be adopted to guide its development and application?

It has been traditional to address questions to do with the introduction of IT in the order in which they appear above. This may be viewed as a 'technology-led' approach. In this book, however, we propose a *psychology-led* approach, since only thus will the technological developments be based on well-understood needs, expectations and effects. Logically, therefore, the approach should start with the questions listed under number 4, that is to consider the possible behavioural and social effects of IT on people and organizations. On the basis of answers to this category of questions, products can be conceived and designed for specific aims (questions 3 and 2). This will ensure that the developments in the technology that are sought (question 1) will be influenced by relevant psychological and behavioural criteria. This is the scheme according to which the main themes of this book are summarized in the following sections.

BEHAVIOURAL OUTCOMES AND PSYCHOLOGICAL ASSUMPTIONS

The importance of behavioural considerations

As Howarth points out in his introductory chapter, it is a mistake (often fatal) to leave experience to reveal the behavioural consequences of different applications of advanced technologies. The design of viable and effective information technology products can only take place when the needs of people who are to use them are incorporated at a very early stage and continue to take an important place throughout the design, construction and operation of the technology. After all, it is people who have to operate the machines which have been developed and who have to interact with them to make them function. It is people who have to accept the system and its intrusion into work and home. It is people who are called upon to adapt their habits to the dictates of IT systems. If people cannot or will not do these things, then IT products will not function adequately.

Other contributions to the book develop this theme. IT systems introduced into organizations without sufficient appreciation of social psychological factors will not function effectively. Ergonomic considerations are crucial to the successful operation of the equipment.

In manufacturing industry, in offices, health care, schools and in the home the effectiveness of IT products and systems depends not only on the effective functioning of sophisticated equipment but, crucially, also on psychological and social factors that relate to the operator and to the organization. In the longer term, the acceptability of IT products will depend on the attitudinal and behavioural consequences that come to be associated with them.

The difficulty of anticipating behavioural outcomes

It would be misleading to assume that behavioural outcomes are easy to anticipate. On the contrary, as Howarth emphasizes at the start of the book, 'common sense' is a notoriously unreliable guide to psychological reactions – far more sophisticated methods and techniques are required. As the contributions to this collection illustrate, a complex range of factors needs to be taken into account for an understanding of people's reactions to IT. These include:

- ☐ The nature of IT systems themselves, how easy they are to use, how intelligible, how functionally adequate.
- ☐ The design of work practices associated with IT, including quality of working life outcomes and the effect of IT on the fabric of the social life in an organization.
- ☐ The organizational outcomes of IT systems, including such factors as the emphasis that has been placed on different functional priorities in their design; the costs and benefits of the systems to different groups; and the effects on power structures and control processes.
- ☐ At the social level, the usability of products and the effects they are found to have on people's customs and culture.

Relevant models and concepts from psychology

A number of psychological concepts and models have been introduced in this collection to make sense of this complex of factors. They may be summarized under two headings. First, the *usability* of IT equipment and its *functionality* have been stressed. To be usable, any piece of equipment in which humans have an input must be designed to be operated by them. Similarly, to be functional IT products need to meet not only technological specifications but also the needs and motivations of potential users. Psychological theories of cognitive, perceptual and motor skills (for example, Chapter 4) are of central relevance to the design process. So also are theories of information processing and decision making (Chapters 5 and 7) and models of attitudes and specific behaviour patterns (for example, Chapters 10 and 11).

Second, it has been emphasized that information technology should not be thought of as mere technological gadgetry. In any particular application it is advisable to think of it as *a pattern of social relationships*. Such an approach can help to explain the variety of reactions that IT products and systems may provoke and make them understandable. There is no simple 'cause and effect' relationship between IT systems and the attitudes, behaviours and organizational structures that come to be associated with them. What is crucial is how those affected come to regard IT products and systems, and the meanings that they come to have. Reactions to new technologies cannot be understood in isolation from the manner of their introduction and the circumstances of their usage. The relevance of psychological approaches to the management of change (Chapter 2), the need for an 'interest group' approach to understanding relative costs and benefits (Chapter 6) and the significance of social changes that IT systems may facilitate or necessitate (for example Chapters 3, 8 and 10) are all stressed within this collection.

ORGANIZING FOR TECHNOLOGICAL DEVELOPMENT AND APPLICATION

Given the importance of behavioural considerations for the success of IT products, close attention is needed to ways in which they can be incorporated into processes of design and introduction. Psychological approaches can be of use in a number of ways, and the methods discussed in this volume are outlined in the following sections.

Designing for usability

Knowledge of human factors can be embodied in guidelines for engineers or in the specification of performance and product standards. Given the importance and complexity of relevant knowledge, however, the direct involvement of psychologists in the design process is likely to be beneficial. The advantages of ergonomic approaches to the design of the operator–machine interface (Chapter 4) are well established, while the value of a 'human factors orientation' is clearly illustrated by the benefits that well-designed information technology products can bring to the disabled (Chapter 8). The advantages of applying psychological techniques to the task of software development (Chapter 5) are also increasingly being accepted. Similarly, psychological approaches are central to the satisfactory development of knowledge-based systems (Chapter 7) and to the design of certain consumer products and services (Chapters 11 and 12). In these and other examples in this book it is made clear that there is no substitute for careful attention to psychological factors and their incorporation within the design process. The benefits of careful empirical evaluation of the success of such an approach (see, for example, Chapters 2, 3 and 5) are also evident.

Understanding the relevance of social psychological factors

Generally, methods to enable social psychological factors to be incorporated within the design and introduction process are less well established than are cognitive, physiological or anthropometric approaches suggested by ergonomists. However, a number of approaches are introduced within this collection that are of direct relevance to design procedures. These include:

- ☐ Analysing pressures within organizations that discourage an emphasis on social psychological factors (see Chapters 2 and 3).
- ☐ Involving end-users in the selection and design of IT products (Chapter 2).
- ☐ Applying motivation theory to jobs built around IT work systems (Chapter 10).
- ☐ Using organizational theory in a pluralist analysis of organizational goals and applying of an 'interest group' approach to evaluation (Chapter 6). Considering the significant changes in conventional patterns of social interaction that can occur as a result of using IT systems (for example, Chapter 9).

Clearly, close evaluation of social outcomes arising from the introduction of IT products is as beneficial as exploring their usability.

GOALS FOR INFORMATION TECHNOLOGY

The functional effectiveness of IT requires that the relevant human goals are defined correctly, in terms both of impact and cost-effectiveness.

In terms of the immediate impact of an IT product or system, relevant goals include:

- ☐ the acceptability and intelligibility of equipment and software
- ☐ relevance and appeal of the product or system to potential consumers
- ☐ longer-term social psychological benefits for users, in terms of quality of work or quality of life outcomes.

Additionally, however, it is clear that the costs and benefits of information technology may not be equitably distributed within society. Relevant considerations here are:

- ☐ The relative costs and benefits to people at different hierarchical levels within organizations that introduce IT systems; 'technology-led' developments, for example, tend to result in reduced opportunities for involvement and decision making at junior levels (Chapters 2 and 3).
- ☐ The relative benefits to different or functional groups within organizations; 'political' considerations between different interest groups can be crucial to the design process of IT systems (Chapter 6).
- ☐ Within society more generally, the costs and benefits of IT are unlikely to be distributed equally. Changes in employment patterns will inevitably result; changes in the requirements for services provision will emerge; changes in the administration and operation of state information and regulatory functions will also take place (Chapter 10).

The relative merits and demerits of such possible outcomes is a matter for debate, of course. What is crucial, however, is that the importance of human-centred criteria for the introduction of information technologies is recognized and that psychological knowledge and methods are used to inform their development and evolution.

POLICY FOR TECHNOLOGICAL DEVELOPMENT

A key message of this volume is that attention must be given to the psychological implications of IT at all stages of policy making. At the start of the postscript we commented on the strong emphasis currently being given to issues of technological development alone. But technological development, on its own, is not enough. With a technology so flexible in its applications as microelectronics it is crucial that priority be given also to its possible behavioural, attitudinal, organizational and social consequences.

A strong commercial thrust will, no doubt, continue to drive the emergence of new products; new and novel applications will continue to be developed. Impressive though microelectronic technologies are, however, the medium is *not* the message. As advanced systems are introduced with increasing frequency and as information systems grow in size and significance, the kinds of approaches that are today associated with the introduction of IT would, if continued, have momentous implications. No matter how innovative and advanced IT products may become, their impact will be constricting if they are not designed for operator usability. Moreover, it is neither sensible nor economical to divorce the design and development processes from an appreciation of the circumstances in which they will be used.

In conclusion, therefore, two messages arise from this volume. First, it is inevitable that considerable problems will emerge if human-centred criteria are not employed to guide the design and implementation of information technology.

Second, a psychology-led orientation to technological development will introduce a clear purpose to the process of technological development. It requires that technologies should be fashioned in ways that are compatible with human characteristics and functioning, and should be applied with regard to personal and social consequences. Examples abound within this volume to demonstrate that in terms of the final acceptability of IT products, their performance and the quality of goods and services they are used to provide, the opportunities for psychology-led approaches are impressive. With the help of psychology, progress towards the coming 'information society' will be made more rapidly and the society that emerges will be more efficient in its operations, more effective in its achievements and more convivial in its organization.

INDEX

Made in United States
Orlando, FL
05 November 2026

90229837R10149